DATE LOANED

CROP QUALITY, STORAGE, AND UTILIZATION

CROP QUALITY,
STORAGE, AND UTILIZATION

C. S. Hoveland, Editor

American Society of Agronomy
Crop Science Society of America
Madison, Wisconsin
1980

FOUNDATIONS FOR MODERN CROP SCIENCE SERIES

Crops and Man   1975
J. R. Harlan
Introduction to Crop Protection   1979
W. B. Ennis, Jr.
Crop Quality, Storage, and Utilization   1980
C. S. Hoveland
Propagation of Crops   In preparation
J. C. DeLeLouche
Crop Breeding   In preparation
D. R. Wood
Physiological Bases for Crop Growth and Development
In preparation
M. B. Tesar
Ecological Bases for Crop Growth and Development
In preparation
W. L. Colville

Matthias Stelly, *Coordinating Editor*
Domenic Fuccillo, *Managing Editor*

Library of Congress Cataloging in Publication Data

Crop Quality, Storage, and Utilization
(Foundations for modern crop science series)
   Includes bibliographies and index.
   1. Field crops. 2. Field crops—Quality. 3. Field crops—Storage. I. Hoveland, C.S. II. Series.
SB185.C76   633.08   80-15859
ISBN 0-89118-035-4

The American Society of Agronomy, Inc. and The Crop Science Society of America
677 S. Segoe Road, Madison, Wisconsin, USA 53711

Printed in the United States of America

FOREWORD

The accelerated pace of research, augmented by sophisticated instrumentation and techniques, and new opinions, imparts to crop science a rapidly changing character as new discoveries replace and/or add to former concepts. New findings force us to reevaluate and often reconstruct the foundations on which crop science rests.

The Teaching Improvement Committee of the Crop Science Society of America identified the urgent need for developing contemporary reading materials aimed at upper level undergraduate college students. A current presentation of the dynamic state of modern crop science is a formidable challenge worthy of the best talents of eminent research and teaching personnel in the field. This task necessitates assembling the most capable representatives of the various disciplines within crop science and bringing them together in teams of writers to prepare a series of publications based on contemporary research. The Crop Science Society of America and the American Society of Agronomy have undertaken this large assignment by selecting more than 100 specialists who will contribute to making the Foundations of Modern Crop Science books a reality.

The authors and editors of this series believe that the new approach taken in organizing subject matter and relating it to current discoveries and new principles will stimulate the interest of students. A single book cannot fulfill the different and changing requirements that must be met in various programs and curricula within our junior and senior colleges. Conversely, the needs of the students and the prerogatives of teachers can be satisfied by well-written, well-illustrated, and relatively inexpensive books planned to encompass those areas that are vital and central to understanding the content, state, and direction of modern crop science. The Foundations for Modern Crop Science books represent the translation of this central theme into volumes that form an integrated series but can be used alone or in any combination desired in support of specific courses.

The most important thing about any book is its authorship. Each book and/or chapter in this series on Foundations for Modern Crop Science is written by a recognized specialist in his discipline. The Crop Science Society of America and the American Society of Agronomy join the Foundations for Modern Crop Science Book Writing Project Committee in extending special acknowledgment and gratitude to the many writers of these books. The series is a tribute to the devotion of many important contributors who, recognizing the need, approach this major project with enthusiasm.

A. W. Burger, chairman
C. D. Dybing
A. A. Hanson
L. H. Smith
M. Stelly

v

PREFACE

What is satisfactory crop quality? Many factors shape our definition of quality for a particular crop. The condition of a crop may make it totally unsuitable for one purpose but highly desirable for another use. Crop quality is dependent on many factors, such as ultimate use of the crop, level of economic development, human and animal preferences, diet, government policies, and the climate in which the crop is grown and stored. This diversity is illustrated by our attitudes toward polished vs. unpolished rice, high vs. low protein grain cultivars, and long vs. short staple length of cotton fibers. Thus, crop quality is interpreted on the basis of physical, nutritional, aesthetic, and other characteristics.

Quality has long been recognized as important in determining the value of crops for specific uses. Plant fiber for cloth and cordage, malting barley for brewing, and wheat for bread and pasta have been evaluated and sold on the basis of quality characteristics. In contrast, the nutritive value of food and feed grains have generally received little consideration until the last several decades. Forage quality has received even less attention than in other crops until recent years. Too often, crop production has been concerned with yield alone rather than yield of a high quality product.

Standards of quality needed for particular uses of the crop have been better defined in recent years with knowledge gained from research. Crop quality improvement during production and maintenance of quality during harvesting and storage are often under control of man. Plant breeding, to select specific quality characteristics, has made substantial advances in fiber, oilseed, and grain crops. In contrast, it is only recently that breeding for improved forage quality has shown the enormous potential in this area. Quality of many crops is greatly affected by the conditions under which they are grown, some of which can be managed by man. Harvesting and storage further affect the quality of crops, sometimes determining whether it has any value at all.

This book deals with three major areas of crop quality: 1) quality requirements and utilization, 2) genetic and environmental effects on quality, and 3) harvesting and storage effects on quality. Principles of crop quality are illustrated by three groups of crops: 1) food and feed grains, oilseeds, 2) fiber crops, and 3) forage crops. The omission of other important crops, such as sugarcane, sugarbeets, manioc, potatoes, rubber, tobacco, cacao, coffee, food legumes, fruits, vegetables, and tea is intentional. The original planning committee felt that the objectives of the book should be to cover principles for some major crops rather than specifics on each particular crop species. Therefore, a limited approach was used.

The authors of these chapters have labored to meet these objectives. This book, developed for students in crop quality, should be useful to students in agronomy, food science, and animal nutrition. It may also serve as a general reference for persons interested in broad technical aspects of crop quality.

Carl S. Hoveland
Auburn, Alabama

CONTRIBUTORS

L. S. Bates, Department of Grain Science and Industry, Kansas State University, Manhattan, KS 66506

R. G. Creech, Agronomy Department, Mississippi State University, Mississippi State, MS 39762

J. E. Freeman, CPC International, Inc., International Plaza, Englewood Cliffs, NJ 07632

L. M. Gourley, Agronomy Department, Mississippi State University, Mississippi State, MS 39762

R. L. Haaland, Agronomy and Soils Department, Auburn University, Auburn, AL 36830

E. G. Heyne, Department of Agronomy, Kansas State University, Manhattan, KS 66506

C. S. Hoveland, Agronomy and Soils Department, Auburn University, Auburn, AL 36830

W. G. Monson, SEA-USDA, Tifton, GA 31794

J. E. Moore, Animal Science Department, Nutrition Laboratory, University of Florida, Gainesville, FL 32611

L. E. Moser, Department of Agronomy, University of Nebraska, Lincoln, NE 68503

H. H. Ramey, SEA-USDA, Cotton Quality Laboratory, University of Tennessee, Knoxville, TN 31796

B. A. Waddle, Department of Agronomy, University of Arkansas, Fayetteville, AR 72701

CONTENTS

Part I. Crop Quality and Utilization

Chapter 1

Food and Feed Grain Crops

R. L. HAALAND

Auburn University
Auburn, Alabama

The quality of food and feed grain crops directly or indirectly affects the majority of the human population and domestic livestock of the world. Quality is an inherited feature of a crop subject to environmental modification. Criteria for quality, however, are rather subjective and vary greatly throughout the world. In developed nations—where diverse diets, including meat protein, are possible—quality criteria are often physical. Factors that affect the ability to process and market a commodity are often given priority. For instance, cereal grains that have clean, disease-free, sound kernels with a high test weight (weight per volume) are considered high quality. Little attention is paid to nutritional aspects, such as biological value of proteins or vitamin and mineral content.

In developing nations, 70 to 90% of the food produced never reaches a commercial market, and diets of one or two items are commonplace. Meat may be a rarity because of economic and regional limitations on animal husbandry or traditional and religious taboos. Growing food is often a struggle for existence so quality criteria are often based simply on dietary preference. Corn (*Zea mays* L.) that makes good tortillas or rice (*Oryza sativa* L.) that sticks together when cooked may be considered good quality. Subsistence agriculture focuses little attention on nutritional quality criteria. Yet, as evidenced by Fig. 1.1, it is in the developing nations that improved nutritional quality in food crops could contribute the most to mankind. This chapter delves into both physical as well as nutritional quality of food and feed grains.

HUMAN NUTRITIONAL REQUIREMENTS

Nutritional needs are based on requirements for energy, amino acids, vitamins, and minerals, all of which are essential for life. Energy is required for basic metabolic processes which maintain the body by functioning of vital organs, building of tissues, and reproduction.

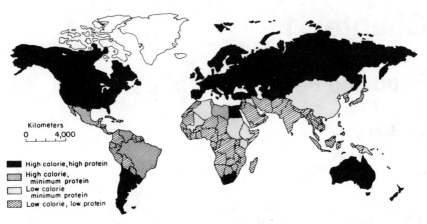

Fig. 1.1—The geography of hunger.

Energy supplied and used by the body is usually expressed as kilo-calories or Calories. The Calorie measures the energy produced by food oxidized in the body. Carbohydrates, fats, and proteins can be utilized by the body as energy sources. Cereal carbohydrates contribute up to 75% of the calories consumed by man in some developing countries, while in other areas, such as the Arctic region, energy is derived mostly from animal fat and protein.

Tissue building is highly dependent on the "building blocks" of life called amino acids. Amino acids are supplied by the proteins of plant and animal tissues. Although proteins are made up of 20 different amino acids, only eight have been established as essential for human beings (Table 1.1). The remaining 12 are synthesized within the body. Several vitamins and minerals are required by man (Table 1.1), some of which are supplied by crop plants, others by animal products.

ANIMAL NUTRITIONAL REQUIREMENTS

Nutritional requirements of animals depend on the digestive system of the animal. All animals require energy for metabolism, growth, and reproduction. Ruminant animals can utilize cellulose (cell wall carbohydrate) as an energy source because the microorganisms of the rumen are capable of breaking down the cellulose to readily oxidizable carbohydrates. However, non-ruminant animals require their energy source in a more digestible form, such as starch, fat, and protein. Non-ruminant animals have requirements for essential amino acids that must be met by their feed supply. Ruminant animals are supplied with essential amino acids by the microbes in the rumen that can use non-organic nitrogen for amino acid biosynthesis. The fatty acids, linoleic, arachidonic, and linolenic are essential for normal body growth in non-ruminant animals. These essential fatty acids are widely distributed

Table 1.1—Nutrients required by man.†

Essential amino acids	Essential vitamins	Essential minerals
Aromatic	Water-soluble	
Phenylalanine	Vitamin B₁ (thiamine)	Calcium
	Vitamin B₂ (riboflavin)	phosphorus
Basic	Niacin	Sulfur
Lysine	Vitamin B₆ (pyridoxine)	Potassium
Histidine	Pantothenic acid	Chlorine
Branched chain	Folacin	Sodium
Isoleucine	Vitamin B₁₂	Magnesium
Leucine	Biotin	Iron
Valine	Choline	Fluorine
Sulfur-containing	Vitamin C (ascorbic acid)	Zinc
Methionine	Fat-soluble	Copper
	Vitamin A (retinol)	Silicon
Other	Vitamin D	Vanadium
Tryptophan	Vitamin E (tocopherol)	Tin
Threonine	Vitamin K (phylloquinone)	Nickel
		Selenium
		Manganese
Arachidonic		Iodine
Linoleic		Molybdenum
Linolenic		Chromium

† From: Scrimshaw and Young (1976).

among fats from feeds such as corn, soybeans [*Glycine max* (L.) Merr.], cottonseed (*Gossypium hirsutum* L.) and peanut (*Arachis hypogea* L.) oils. Animal nutrition is covered in more detail in chapter 3.

CEREAL PRODUCTS, UTILIZATION, AND QUALITY

Cereal grains are major contributors to human nutrition throughout much of the world. Most grain is consumed after some type of milling followed by cooking. The type and quality of products obtained from cereal grains depend on the type and quality of the grain used and its processing.

Wheat Dry Milling Food Products and Utilization

Wheat (*Triticum aestivum* L.) has been cultivated and used as a food source for 6,000 to 8,000 years. Man developed milling and baking techniques from 4,000 to 3,000 B.C. in both Mesopotamia and Egypt. Bread wheat is thought to have gradually evolved during this same time period (Cotton and Ponte, 1973).

Wheat was first used as a thick porridge made from ground grain and water; however, it was discovered that if the mixture was allowed to dry, a product more easily stored and transported was formed; thus the first unleavened bread. Leavened breads probably were accidently discovered when yeast cells fell into a wheat porridge mixture, resulting in a lighter textured bread when baked than the unleavened bread.

Table 1.2—Percent protein in wheat used for different products.†

Product	Protein	Type of wheat
Noodles	12.5 to 15	Durum or soft wheat
Macaroni	12.5 to 15	Durum
White bread	11 to 14	Hard red spring or winter
All purpose flour	9.5 to 12	Soft red or white
Crackers	9.5 to 11	Soft red or white
Pie crust	8 to 10.5	Soft red or white
Donuts	8 to 10.5	Soft red or white
Cookies	8 to 9.5	Soft red or white
Cake	8 to 9.5	Soft red or white

† From: Martin et al. (1976).

As civilization advanced, so did the baking industry. Commercial bakeries were part of ancient Greek and Roman societies (Cotton and Ponte, 1973). Today the baking industry is world-wide, producing many diverse products.

The majority of the wheat consumed today is in the form of yeast leavened breads, including white pan bread, rolls and buns of various configuration, white hearth breads—both French and Italian types—and bread with varying amounts of whole wheat flour. These products are usually made with flour that has been produced by dry milling high protein, hard red spring and hard red winter wheat grain (Table 1.2).

Many other types of baked goods produced by the baking industry utilize wheat flour. These include a wide variety of pastries, cookies, crackers, cakes, and pretzels. Flour from low protein soft red winter wheat and soft white wheat is used for these products (Fig. 1.2 and Table 1.2).

Paste products include both pasta and noodles. Pasta refers to products such as macaroni and spaghetti which are made from flour from durum wheat and formed by extrusion of a dough through some type of die. Noodles may be made from durum wheat but are often made from soft wheats by sheeting and cutting the dough. The use of wheat for paste products is more widespread throughout the world today than is its use in bread. This is probably because paste products are easier to make and when dried can be stored for relatively long periods of time (Matsuo, 1975).

CORN AND SORGHUM PRODUCTS AND UTILIZATION

Corn, native to Mexico and Central America, was the major food source of the great Indian civilizations of the Western Hemisphere. Transported to Europe by early New World explorers, corn soon spread throughout Europe, Asia, and Africa. Grinding the mature corn grain between stones, a primitive dry milling, was most frequently used to make corn more palatable.

Dry milled corn products most commonly used for human food are grits, meal, and flour. Corn grits are produced by grinding and sifting white or yellow corn after the bran and germ have been removed.

Fig. 1.2—A variety of products are produced from wheat flour.

Grits, a popular food in the southeastern U.S., are simply cooked in boiling water and used as a side dish. Grits are also used as a carbohydrate source in the brewing industry and may also be flaked to produce corn flakes, a common breakfast food.

Corn meal, bolted corn meal, and degermed meal are produced by dry milling. Meal is ground to a finer texture than grits. Whole meal is made by grinding the whole corn grain. Since whole meal contains fat from the germ it becomes rancid with prolonged storage which limits its utilization. Bolted meal is produced by sifting or bolting whole corn meal which removes some of the bran and germ particles, leaving a brighter meal. Because some of the germ remains in the meal, storage may be a problem. Grinding corn that has had the germ portion of the kernel removed results in degermed meal which has better color and fewer specks than other types of meal. Degermed corn meal is more popular than other corn meals because it has better color and storage attributes because of its lower fat, germ, and bran content (Table 1.3). Corn meal is utilized in various types of corn bread, muffins, mush, and porridge.

Corn flour is produced by grinding clean white or yellow corn and removing the bran, germ, grits, and meal. Corn flour is less granular than grits or meal. Many types of baking mixes, pancake mixes, cooking mixes, and snack foods contain corn flour. Corn flour is also used in meat processing as a sausage or meat binder. Tortillas, a major food in

Table 1.3—General quality attributes of dry milled corn products.†

Quality component	Grits	Whole meal	Bolted meal	Degermed meal	Flour
			%		
Moisture	12.0	12.0	12.0	12.0	12.0
Protein	8.7	9.2	9.0	7.9	7.8
Fat	0.8	3.9	3.4	1.2	2.6
Carbohydrate	78.1	73.7	74.5	78.4	76.8
Fiber	0.4	1.6	1.0	0.6	0.7
Ash	0.4	1.2	1.1	0.5	0.8

† From: Brockington (1970).

Table 1.4—Parts of a wheat kernel as percent of weight of the whole kernel and general quality attributes of each part.†

Component	Pericarp	Aleurone layer	Starchy endosperm	Germ
Percent of weight of kernel	9	8	80	3
Ash, %	3	16	0.5	5
Protein, %	5	18	10	26
Lipid, %	1	9	1	10
Crude fiber, %	21	7	0.5	3

† From: Pomeranz (1973).

Latin America, are prepared with flour milled from corn that has been steeped in alkali.

About 75% of the world's sorghum (*Sorghum bicolor* L.) crop is used for human consumption, making it the third most important food grain in the world (Hahn, 1970). As with corn, sorghum dry milling produces grits, meal, and flour. Utilization of sorghum products is similar to corn except that it is used more commonly in Asian and African countries.

Wheat Dry Milling Quality Criteria

Quality criteria important to the wheat miller as well as the grower would include sound, plump kernels having good color, absence of diseased kernels or foreign material, and high density or test weight (weight per volume). These criteria have been defined by the U.S. Department of Agriculture to develop differential grades within each species of cereal grain that will facilitate marketing, storage, financing, and futures trading (Martin, 1976). Official U.S. grain grades, however, do not include any criteria for nutritional quality. Canadian wheat grades guarantee a minimum protein percentage.

Other factors may also be important to the miller and grower. A kernel moisture content of 12 to 13% results in better storage characteristics and less spoilage than higher moisture levels. Thus, the grain will be in good condition when it reaches the miller. Protein content is a quality criterion of interest to the miller because it will affect the baking performance of the flour; of interest to the grower because a price premium is often paid for specific protein concentrations in the grain.

The starchy endosperm of the wheat kernel, from which white flour is derived, is surrounded by a tough, cellulosic, outer coat called bran (Fig. 1.3). Before wheat is milled, water is added to the grain in a process called tempering or conditioning to toughen the bran and mellow the endosperm. Because of the difference in friability of the two kernel components after tempering, the bran can be thoroughly removed during milling. General quality characteristics of the various parts of the wheat kernel are given in Table 1.4.

Wheat is dry milled by grinding the tempered wheat kernels between a pair of corrugated break rollers rotating against each other. This breaks the kernel and loosens the bran from the remainder of the kernel. The broken particles are sifted through a series of screens to

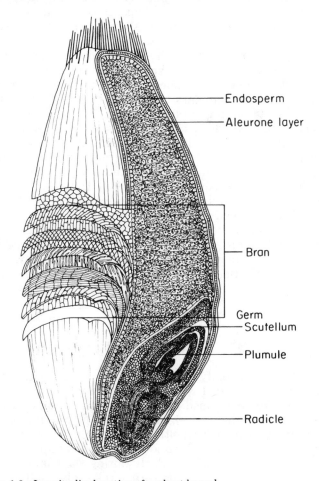

Fig. 1.3—Longitudinal section of a wheat kernel.

separate various size particles. The coarse material is returned to the next successive break roll for further grinding while the fine particles are sent through a purifier which separates small bran particles from granular flour with air currents and screens. Four to five break rolls and siftings are necessary to reduce the wheat particles to granular flour or middlings nearly free of bran. The middlings are further ground on a series of smooth rollers to produce the fine textured flours used in baking.

Milling extraction, the percentage of the wheat contained in the flour after milling, may vary from about 75% (normal white flour) to 100% (whole wheat flour). Components excluded from the flour with less than 100% extraction, bran and germ particles, are used for animal feed. Chemical composition of the flour varies slightly with milling extraction (Table 1.5).

Flour that has been freshly milled will not produce good commercial quality bread and crackers. Naturally aged flour will counteract this quality problem. However, natural aging is often not uniform and long periods of storage may lead to spoilage or insect problems. Aging is hastened by adding oxidizing agents to the flour, giving it improved dough handling properties as well as improved loaf volume, texture, and freshness retention in bread. Bleaching agents are often added to commercial flour to make it white. Fresh ground flour contains pigments that make the finished products less attractive for marketing. Malted barley (*Hordeum vulgare* L.) or malted wheat may be added to provide fermentable sugars and enzymes to hydrolyze starch. Unlike bread wheat flour, durum milled products decrease in quality with aging. The pigments of the durum wheat that give macaroni products their distinctive yellowish color are destroyed by oxidation (Walsh and Gilles, 1973). If durum semolina or flour must be stored, it should be done at low temperature and humidity.

Vitamins and minerals are commonly added to "enrich" the flour, which actually restores the vitamins and minerals that were removed during the milling process. Since the dry miller's end product, flour, is one of the main ingredients of the baking industry, dough and baking quality of the flour are important. Many flour mills have dough mixing and baking facilities to determine the quality of the end products of their flours.

Bread Baking and Quality Criteria

Bread is usually made by mixing flour, yeast, salt, sugar, shortening, and yeast food with water. The ingredients are mixed until the dough has developed desired handling properties. The dough is allowed to ferment (rise), divided, and placed into pans. Panned bread undergoes a final fermentation called proofing which results in additional CO_2 for leavening and modified dough proteins. Proofed dough is baked at about 230 C. Early during the baking process a rapid expansion of the dough (oven spring) takes place followed by a gelatinization of starch and coagulation of proteins, which results in the familiar

Table 1.5—Milling extractions of wheat and general quality attributes of each part.†

Component	Milling extraction		
		%	
Products			
Flour %	75	85	100
Animal feed %	25	15	0
Percent of flour			
Ash	0.5	1	1.5
Protein	11	12	12
Lipid	1	1.5	12
Crude fiber	0.5	0.5	2

† From: Pomeranz (1973).

rigid structure of bread. Various chemical reactions during baking cause the crust to brown and flavor to develop. A reaction between reducing sugars and amines causes the browning. Flavor components produced during this reaction diffuse toward the interior of the loaf as it cools.

Bread produced from some cultivars of wheat may have excellent loaf volume, crumb structure, and shelf life while other cultivars may produce inferior, dense loaves unacceptable to the consumer. These differences may be due to different grain protein content and/or test weight which can be affected by N fertilization rate, temperature at which the crop is grown, amount of rainfall, and the cultivar that is being grown. Baking tests are conducted to determine what quality of bread can be expected from a certain batch of flour.

Baking tests are relatively accurate. The tests are done on a microscale using from 10 to 100 g of flour. Ingredients other than flour are added in excess so they do not become limiting factors. Loaf volume is the most important measurement because it is highly correlated with dough handling properties, crumb texture, freshness retention, and technological versatility (Pomeranz, 1973).

Quality attributes of bread depend on protein and starch content of the flour. Interactions between protein and starch formed during dough mixing and baking are also important.

Gluten, a storage protein complex in wheat, imparts many of the physical quality attributes to bread. Gluten contains large amounts of the amino acid glutamine, up to 32% or more of the total amino acids (Pomeranz, 1973). Gluten is low in the essential amino acid lysine; thus the protein responsible for good physical quality in bread does not have good nutritional quality.

Gluten contains two major protein components, gliadin and glutenin, as well as about 5 to 10% lipid and 10 to 15% carbohydrate (Fig. 1.4). During dough mixing, gluten is hydrated and forms a matrix in which starch, yeast, and other components are imbedded. The gluten framework of dough is responsible for gas retention, which results in leavened products.

Gluten is elastic and cohesive when hydrated. The glutenin fraction controls the elasticity because it is tough and not easily stretched.

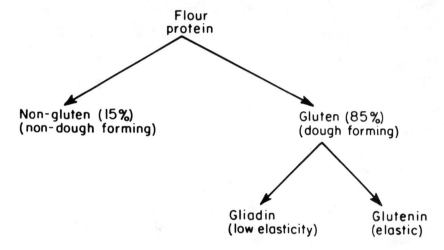

Fig. 1.4—Components of bread flour.

The gliadin fraction is less cohesive and can be stretched. Fractionation studies show that gliadin controls loaf volume and glutenin controls mixing time and dough development (Pomeranz, 1973).

The viscoelastic properties of gluten have been attributed to sulfhydryl groups and to disulfide, hydrogen, hydrophobic, and ionic bonding (Fig. 1.5). Disulfide bonds are probably of most importance because reduction of them causes complete loss of dough cohesion (Pomeraz, 1973).

Glycolipids are bound to glutenin by hydrophobic bonds and to gliadin by hydrogen. The increase in loaf volume during bread-making has been attributed in part to interactions between glycolipids and gliadin (Pomeranz, 1973).

The ability of wheat endosperm proteins to form an elastic dough and retain gas has played a greater role in the acceptance of wheat in human diets than has nutritive value.

Starch, which may make up to 85% of the flour, plays several roles in breadmaking. Starch granules are fractured during milling. These fractured granules are the substrate for amylase, which converts starch to fermentable sugars. Intact granules are slowly converted while severely fractured granules are rapidly converted. Starch gelatinization is important in development of crumb stretching. Degradation of starch will contribute to deterioration of freshness.

Quality Attributes to Soft Wheat Product

Many baked goods, such as cakes and cookies, do not require the extensive leavening process that bread does. Thus, wheats with a lower percentage of grain protein, such as soft red or soft white wheat, can be used for these products. Cookie flour as compared to bread flour

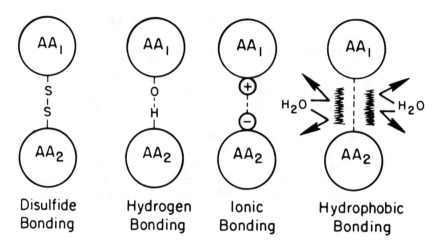

Fig. 1.5—Chemical bonds formed between amino acids (AA$_1$ and AA$_2$) in bread dough proteins.

has lower protein content (8 to 11%), lower water absorption capacity, less starch damage, finer granulation, and produces greater cookie diameter and has more mellow gluten properties. The quality criteria for cakes, cookies, and crackers have been developed to satisfy the baker's and merchandiser's needs rather than on sound nutritional requirements.

Because cookies and crackers have a relatively long shelf life and are popular with children, it may be possible to subtly improve the nutritional intake of youngsters by improving the nutritional value of cookies and crackers. This may be done by using enriched flour or by using supplemental high lysine flours, such as soybean flour.

Quality Attributes of Durum Wheat Products

Pasta products have unique quality requirements because they are made from semolina rather than flour. Semolina is more granular than flour and must meet specific size requirements. Separation is usually done by passing semolina through a series of sieves.

Color is an important quality attribute of pasta products, with a yellow color preferred. Bran particles detract from color and appearance so the number of specks is monitored in semolina. From 10 to 20 specks per 65 cm^2 are normal for good quality semolina (Walsh and Gilles, 1973).

Cooking quality of pasta products depends on how the products stand in boiling water. The amount of water absorbed, the loss of solids to the cooking water, and firmness of the cooked product are important components of quality.

Nutritional Quality of Wheat Products

Most of the quality criteria used by the milling and baking industries have been based on industry needs rather than on nutritional needs of human beings. Nutritional considerations play a lesser role in defining wheat quality in the USA because supplements are added to milled products and dietary intake often includes well-balanced animal protein.

However, in developing countries, diets are often not varied and supplementation of milled products is not normally practiced because the grain is processed by the people who grow it. In these areas, cereal grains must be developed that contain more lysine and tryptophan or be supplemented with foods such as soybeans or other food legumes which contain adequate lysine and tryptophan.

Corn Dry Milling Quality

Corn is dry milled to make the kernels more palatable and the nutrients contained in the grain more digestible. Table 1.6 gives the general chemical composition of dent corn.

There are two methods of dry milling corn. The non-degermed method grinds corn (yellow or white dent) into meal without removing the germ portion of the kernel (Fig. 1.6). Non-degermed corn meal may be bolted (sieved) to remove some of the bran and germ particles. Meal produced this way often has an oily flavor, is soft to the touch and may contain dark specks due to remaining bran particles. Long periods of storage of non-degermed meal are difficult because the high oil content of the meal results in rancidity (Brekke, 1970).

Degerming removes the bran, germ, and tip cap from the kernels before the grain is ground. Toughening the bran by controlled addition of water makes separation of the kernel components easier. Because the germ contains about 35% oil it is often recovered for oil extraction (Brekke, 1970).

Several physical quality criteria are important to the dry miller. Naturally dried corn is preferred over artificially dried corn because the latter often contains cracks that lead to broken kernels and subsequent poor milling yield. Large, sound unbroken kernels with a moisture content of 15% and a test weight of 56 lb/bu (25 kg/0.87 hl) are preferred because small and broken kernels are lost during premilling cleaning, and broken kernels do not temper as uniformly as sound kernels. Only one color of corn is milled at a time. The grain must be free from microorganisms that may produce toxic compounds (Bekke, 1970). Aflatoxin, formed by the fungus *Aspergillus flavus* is highly toxic to human beings as well as livestock.

Fresh corn is better for dry milling than older corn. Fat migrates from the germ into the endosperm with age. This process may lead to end products with widely different oil contents which could affect

Table 1.6—Approximate chemical composition of dent corn.†

Component	Amount	
	%	
Water	13.5	
Protein	10.0	
Oil	4.0	
Starch	61.0 ⎫	
Sugars	7.4 ⎬ Carbohydrates	
Crude Fiber	2.3 ⎭	
Ash	1.4	
Miscellaneous	0.4	

† From: Martin et al. (1976).

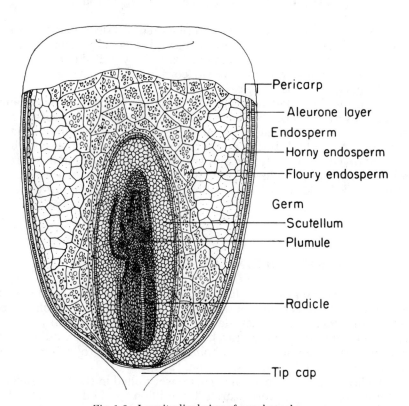

Fig. 1.6—Longitudinal view of corn kernel.

physical and nutritional quality as well as storage properties (Brekke, 1970).

Depending on the end product desired, the corn should vary in percent horny (hard starch) versus floury (soft starch) endosperm (Table 1.7). Grits require more horny endosperm to retain their granular structure while flour requires a finer granulation so the softer floury endosperm will yield more and better quality flour (Brekke, 1970).

Table 1.7—Corn endosperm configuration for various dry milled products.†

Product	Endosperm type	
	Horny	Floury
	%	
Grits	70	30
Meal	45	55
Flour	20	80

† Data from: Brekke (1970).

Nutritional Quality of Corn Dry Milled Products

The dry milling industry has emphasized physical quality criteria over nutritional criteria for many years. Only a small portion of the per capita dietary intake in the USA consists of dry milled corn products. However, in some parts of the world, corn is a major part of the diet so nutritional quality criteria are important.

Corn supplies adequate amounts of energy in the form of carbohydrates; however, it does not have a nutritionally balanced protein. It is deficient in the essential amino acids lysine and tryptophan. The incorporation of a high lysine mutant, Opaque-2, into normal corn lines has improved both the lysine and tryptophan levels. Similar improvements have been made in grain sorghum which is processed and utilized much like corn. This subject is discussed in more detail in chapter 4.

WET MILLING OF CEREAL GRAINS

Several cereal grains including corn, sorghum, wheat, and rice are wet milled to extract pure starch from the endosperm. Starch extraction from cereals is not a new development. As long as 3,000 years ago the Egyptians were using starch in cosmetics and as an adhesive to hold papyrus together. Roman nobility used starch to stiffen clothes during the Roman Empire (Anderson, 1974).

Today over 5 billion kg of corn are wet milled in the USA yielding about 2.7 billion kg of corn starch, 160 million kg of oil, and 1.4 billion kg of assorted feed products. The other cereals are not processed for starch in as large quantities as corn because they are used more for food purposes (Anderson, 1970).

A typical wet milling plant steeps the grain in water, removes the germ portion of the grain for oil extraction, and separates the protein and fiber from the starch (Fig. 1.7). The end products of wet milling include purified starch, oil, bran, protein or gluten, oil meal, and other assorted fat products.

The total starch content varies among cereal grains (Table 1.8) and, depending on cultivar, will vary within the different cereal

Table 1.8—Approximate starch content of cereal grains.†

Cereal	Starch	
	Whole grain	Endosperm
	%	
Corn	72	88
Wheat	65	79
Rice	81	90
Sorghum	74	83
Barley	60	77
Oat	45	70

† From: Medcalf (1973).

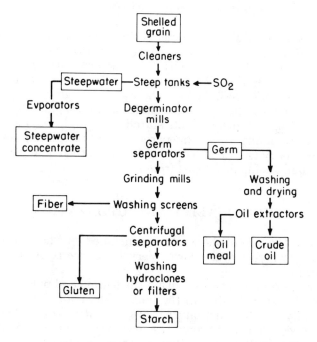

Fig. 1.7—Corn wet-milling flow diagram.

species. Starch granules vary in size and shape among species (Fig. 1.8). Size differences alter rates of gelatinization and hydrolyzation. The quality requirements of starch depends on how it is to be used. Starch to be used in cooking must be easily gelatinized, while starch that is to be converted to sugar or syrup must be easily hydrolyzed. More than half of the starch extracted from corn is converted into corn syrup or glucose and one-third is converted into corn sugar or dextrose.

Starch is composed of repeating glucose units. Two different bonding patterns result in two different long-chain polymers in starch.

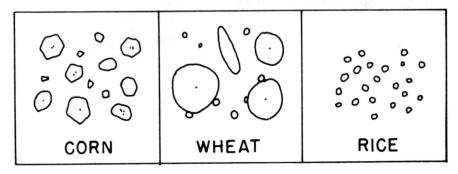

Fig. 1.8—Microscopic appearance of various granular starches.

Amylose is a linear-chain polymer while amylopectin is a branched-chain polymer (see structures in chapter 5). Cultivars of several cereal grains have been developed with varying amounts of amylose and amylopectin (Table 1.9). The different starch components fulfill different specialized roles in industry. The waxy or high amylopectin varieties are commonly used in instant pudding mixes, glues and other materials requiring branched-chain starch. The high amylose corn has been used to make thin transparent film, resembling cellophane. Starch chemistry is still not completely understood, and new products are continually being developed from starch.

RICE QUALITY AND UTILIZATION

Most of the rice in the world is consumed as whole kernel rice in the country where it was grown. Very little rice is used for flour or baked goods.

The rice grain is enclosed within the lemma and palea, which form a rigid hull (Fig. 1.9). Rice in the hull, called paddy or rough rice, is mechanically dehulled to produce brown rice. Rice grains that have the bran attached are called brown rice and may be milled to remove the bran and germ. Milled rice is often run through a brushing operation to further remove bran particles along with some endosperm tissue. This processing decreases the nutritional value of the rice grain. The resulting rice grains are usually white or light tan in color. The screenings from the polishing operation are called rice polish and are often used in animal feed.

About one-fourth of the world's paddy rice crop is parboiled before it is milled (Gariboldi, 1973). Parboiling consists of steeping and/or steaming the paddy rice by various methods. Parboiled rice breaks less when it is milled and tends to store better than non-parboiled rice. Parboiling also moves the water-soluble B vitamins from the bran and hull into the endosperm. It was observed in the early 1900's that people who consumed parboiled rice seldom were affected by beriberi, a disease caused by the lack of vitamin B_1.

Table 1.9—Amylose and amylopectin content of several types of cereal grains.†

Cereal	Amylose	Amylopectin
	%	
Corn	24	76
Waxy corn	1	99
High amylose corn	75	25
Wheat	25	75
Rice	18	82
Waxy rice	1	99
Sorghum	25	75
Waxy sorghum	1	99

† From: Medcalf (1973).

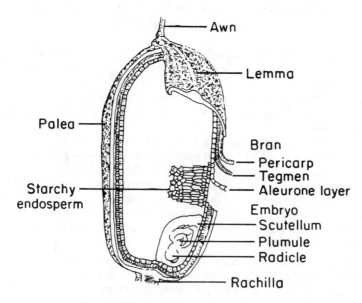

Fig. 1.9—Rice grain structure.

The texture and gloss of cooked rice are determined by the amylose:amylopectin ratio (Gariboldi, 1973). The long grain rice cultivars, commonly grown in the USA, are high in amylose. Long grain rice, when cooked, has a light fluffy texture preferred in many Western countries. Asians prefer the short and medium grain rice cultivars which contain less amylose than the long grain varieties and have a sticky texture when cooked which makes it easier to eat with chopsticks.

Protein content of rice varies from about 7 to 15% depending on cultivars and environment. The main protein fraction is called oryzenin but it does not have the CO_2 retention characteristics of wheat gluten. Like other cereal grains, rice protein is not well balanced because it is low in the essential amino acid lysine. In many Asian countries this becomes an important nutritional limitation because rice may comprise up to 80% of dietary intake.

Table 1.10—Average composition of cereal grains.†

Component	Brown rice	HRW‡ wheat	Corn	Sorghum	Millet	Pearled barley	Oats	Rye
Moisture, %	12.0	12.5	13.8	11.0	11.0	11.1	8.3	11.0
Calories/100 g	360	330	348	332	327	349	390	334
Protein, %	7.5	12.3	8.9	11.0	9.9	8.2	14.2	12.1
Fat, %	1.9	1.8	3.9	3.3	2.9	1.0	7.4	1.7
N-free extract, %	77.4	71.7	72.2	73.0	72.9	78.8	68.2	73.4
Fiber, %	0.9	2.3	2.0	1.7	3.2	0.5	1.2	2.0
Ash, %	1.2	1.7	1.2	1.7	2.5	0.9	1.9	1.8

† From: Adair (1972).
‡ HRW = Hard Red Winter Wheat.

COMPARATIVE NUTRITIONAL VALUE OF CEREAL GRAINS

All of the cereal grains have a relatively high caloric value (Table 1.10). In humans and other monogastric animals, caloric intake must be accompanied by adequate protein intake or protein deficiencies will develop. Both protein content of the cereal as well as the amino acid composition of the protein are important in meeting nutritional requirements.

The biological value of a protein depends on content of essential amino acids in relation to the requirement for these amino acids by the species of animal that will use the protein. Table 1.11 gives the amino acid requirements for humans. The essential amino acid that is most deficient in a protein is called the first limiting amino acid. Table 1.12 shows the amino acid composition of cereal grains. In addition to essential amino acids, animals require a non-specific source of N to synthesize non-essential amino acids.

The Protein Efficiency Ratio (PER) is one of the most widely accepted procedures for determining protein quality. Rats are used for bioassay of a protein. The PER is calculated as:

$$PER = \frac{body\ weight\ gain\ in\ grams}{protein\ consumed\ in\ grams}$$

The assay must be highly standardized as to size, age, sex, and environment to be acceptable. Amino acid availability may be reduced by poor digestibility or excessive cooking during preparation. Reactions between free amino groups with sugars, aldehydes, and fatty acids that occur when proteins are heated with other foods may decrease availability of amino acids.

Normal cereal grains do not have nutritionally balanced proteins. The first limiting amino acid in most of them is lysine. Tryptophan may also be limiting. In countries where cereal are the mainstay diets, protein nutrition is a widespread problem. New cultivars of corn, sorghum and barley have been developed that are higher in lysine than

Table 1.11—Amino acid requirements for human beings.†

Amino acid	Infant	Requirements 10–12 years	Adult
		mg/g of protein	
Histidine	14.0	--	--
Isoleucine	35.0	37.0	18.0
Leucine	80.0	56.0	25.0
Lysine	52.0	75.0	22.0
Methionine + Cystine	29.0	34.0	24.0
Phenylalanine + Tryrosine	63.0	34.0	25.0
Threonine	44.0	44.0	13.0
Tryptophan	8.5	4.6	6.5
Valine	47.0	41.0	18.0

† From: World Health Organization (1973).

Table 1.12—Amino acid composition of several cereal grains.†

Amino acid‡	Wheat	Rye	Corn	Barley	Oats	Rice	Sorghum
				%			
Arginine	0.80	0.53	0.51	0.60	0.80	0.51	0.40
Cystine	0.20	0.18	0.10	0.20	0.20	0.10	0.20
Histidine	0.30	0.27	0.20	0.30	0.20	0.10	0.30
Isoleucine	0.60	0.53	0.51	0.60	0.60	0.40	0.60
Leucine	1.00	0.71	1.11	0.90	1.00	0.60	1.60
Lysine	0.50	0.51	0.20	0.60	0.40	0.30	0.30
Methionine	0.20	0.20	0.10	0.20	0.20	0.20	0.10
Phenylalanine	0.70	0.70	0.51	0.70	0.70	0.40	0.51
Threonine	0.40	0.40	0.40	0.40	0.40	0.30	0.30
Tryptophan	0.20	0.10	0.10	0.20	0.20	0.10	0.10
Tyrosine	0.51	0.30	0.50	0.40	0.60	0.70	0.40
Valine	0.60	0.70	0.40	0.70	0.70	0.51	0.60

† From: National Academy of Sciences (1964).
‡ Values reported on moisture-free basis.

normal cultivars. High lysine wheat and rice cultivars are yet to be developed. Nutritionally balanced cereal proteins would have far reaching benefits for developing nations.

MALTING AND BREWING

Malting cereal grains and fermenting the malt to produce beer has been a part of man's culture since prehistoric times. Barley, wheat, rye (*Secale cereale* L.), sorghum, corn, and rice are malted throughout the world. In the USA and many western European countries, barley is the main crop that is used for malting, because it has hulls that remain attached during malting which results in more uniform germination of the seed. The hulls also aid filtration in the brewing process.

Malting is controlled, limited germination of grain used to activate and synthesize enzyme systems. The enzymes ∝-amylase and β-amylase are produced when grains germinate. Starch is most efficiently hydrolyzed to fermentable sugars when both enzymes are present.

Several factors are used to determine malting quality of barley. Some cultivars produce malt more efficiently than others. Mixtures of cultivars are undesirable. Sound, plump kernels, free of disease, and germination above 95% are important malting quality criteria.

Preferred protein levels range from 9.0% in Western six-row barley to 12.5% in Midwestern six-row barley. Twelve percent is the preferred protein level in Western two-row varieties (Peterson and Foster, 1973). Excessive protein content may decrease uniformity of germination and decrease the amount of soluble material that can be extracted from malt. Enzymatic activity, as measured by diastatic power and $\propto$-amylase in malt, increases with increased protein content. If enzymatic activity is too high, beer quality will decline.

Beer is made by extracting soluble material from malt and adding hops and yeast to produce a fermented, hopped, malted beverage. Approximately 20 kg of material are used to produce a barrel (116 liters) of beer: 13 kg malt, 4.5 kg corn grits, 2 kg rice grits in addition to a small amount of hops and yeast (Pomeranz, 1972). Corn and rice grits are used as brewing adjuncts to increase the carbohydrate content available for fermentation. Both rice and corn result in lighter colored, less filling beer than those that do not contain rice or corn.

FERMENTATION AND DISTILLING

In addition to being used in the production of malt, cereal grains are used to produce distilled spirits such as whiskey, gin, vodka, and grain alcohol (ethyl alcohol). Corn, rye, and barley are most commonly used by distillers. Wheat and sorghum are used to a lesser extent.

Some distilled spirits have particular grain requirements while others do not. Vodka and gin can be produced from any grain. The Russians originally made vodka from potatoes (*Solanum tuberosum* L.). Gin has a characteristic flavor and aroma developed by steeping juniper berries in the alcohol. There are several types of whiskey, each requiring a particular grain. Boubon is distilled from a mash of grain containing at least 51% corn. Corn whiskey utilizes at least 80% corn, while rye whiskey must contain at least 51% rye as a cereal mash component. Scotch whiskey is made only in Scotland from a blend of grain alcohol and whiskey produced from malted barley. The distinctive smokey flavor and aroma of Scotch come from drying the malted barley over peat fires. Irish whiskey is produced only in Ireland from a blend of barley malt whiskey and grain whiskey. The malt is dried in coal-fired kilns, however, so it does not have the smokey characteristics of Scotch (Cotton, 1970). In all cases, the grain is used as the carbohydrate source for fermentation.

Cereal grains are also used for other industrial fermentations. The growth of specialized yeasts, molds, or bacteria on carbohydrate-rich substrates under controlled conditions yield various alcohols, organic acids, amino acids, antibiotics, enzymes, pigments, polysaccharides, and vitamins.

FEED GRAINS

The starchy endosperm of cereal grains is an excellent source of carbohydrate and as such can be used as energy for livestock. Cereal grains, depending on how they are processed, may be much more digestible than most forages.

Grain is fed to nearly all classes of livestock, but is particularly important as an energy source in poultry and swine feeds. Cattle that are fattened in feedlots also consume large quantities of grain.

Several types of grain, including corn, sorghum, barley, and oats are often fed as whole grain or with a minimum of processing—coarse grinding in a hammer mill or steam rolling of the whole grain. This processing often increases palatability and extent of digestion. Man consumes most of the wheat and rice produced in the world. However, milling byproducts such as bran, germ, and endosperm particles are utilized as animal feed. Brewing byproducts, particularly spent grains (brewer's grain), are also used for animal feed.

Most cereals are fed in mixed rations along with protein supplements. Cereals are easily digested and move rapidly through the digestive tract of animals. With ruminant animals a roughage such as hay or silage should be added to the ration to slow the passage of the feed so it can be efficiently utilized. A thorough discussion of animal nutrition is given in chapter 3.

The amount of cereal grain used for livestock feed varies considerably in different parts of the world. Developing nations in Africa and Asia use little grain for animal feed. The USA, Canada, and nations in western Europe depend on animal protein in their diets so they use much more grain for feed than do the developing nations (Fig. 1.10). Most of the corn and sorghum produced in the USA is used for livestock feed. However these are major food crops in several South American, African, and Asian countries.

OILSEED PRODUCTS, UTILIZATION, AND QUALITY

Oils have been extracted from plants for centuries for use in food preparation, medicines, and various industrial processes. Many plants were known to contain oil but geographical area governed which ones were predominantly used. For example, in the dry Mediterranean area, olive (*Olea europaea* L.) oil was the major oil. Flax (*Linum usitatissimum* L.) or linseed was used in southern Europe for its fiber as well as oil. Sesame (*Sesamum indicum* L.), a particularly rich source of oil, was probably one of the first oils used in the tropics and the Orient.

In Russia during the 1700's the Orthodox Church had laid down strict restrictions on eating oily foods on certain Holy days. Because the newly introduced sunflower (*Helianthus annuus* L.) was not on the prohibited list, it gained quick prominence as suitable vegetable oil. Today sunflower is the most important edible oil crop in the USSR (Heiser, 1976).

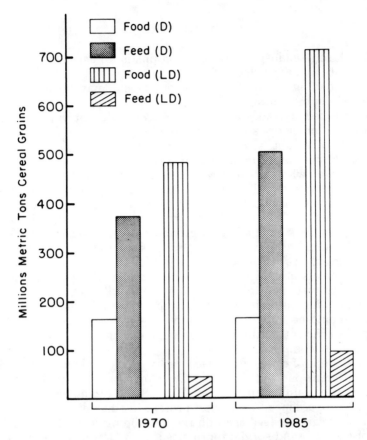

Fig. 1.10—Cereal grain consumption for 1970 and expected in 1985 for developed (D) and less developed (LD) countries.

The more important vegetable oils produced in the world today include soybean, peanut, cottonseed, sunflower, rapeseed (*Brassica napus* L.) and oil palm (*Elaecis guineensis*) (Table 1.13).

Vegetable Oil Utilization

Plant-derived oils, depending on their source, have many diverse uses. As mentioned earlier, oils (fats) are an extremely rich source of energy, but they also are important in the human diet because they supply three essential fatty acids, arachidonic, linoleic, and linolenic. Vitamins A, K, D, and E are soluble in oil, so oils are carriers of some important vitamins.

Fats and oils are used not only to cook food but in food preparation to improve flavor and palatability. Vegetable oil products commonly used for food purposes include cooking and salad oils, vegetable fats or shortening, artificial dairy products such as margarine, and a number of spreads, such as mayonnaise.

Table 1.13—Characteristics of several common oil seeds and vegetable oils.†

Crop	Botanical name	Oil content	Iodine number‡	World production, oil, 1973
		%		1,000 metric tons
Drying oil				
Flax	*Linum usitatissimum* L.	35–45	170–195	790
Tung	*Aleurites fordii* L.	40–58	160–170	140
Safflower	*Carthamus tinctorius* L.	24–36	140–150	285
Semidrying oil				
Soybean	*Glycine max* L. Merr.	17–18	115–140	7,170
Sunflower	*Helianthus annuus* L.	29–35	120–135	3,730
Corn (germ)	*Zea mays* L.	50–57	115–130	305
Cottonseed	*Gossypium hirsutum* L.	15–25	100–116	2,780
Rapeseed	*Brassica napus* L.	33–45	96–106	2,480
Nondrying oil				
Sesame	*Sesamum indicum* L.	52–57	104–118	605
Peanut	*Arachis hypogeae* L.	47–50	92–100	3,020
Castorbean	*Ricinus communis* L.	35–55	82–90	365
Coconut	*Cocos nucifera* L.	67–70	8–12	2,675
Olive	*Olea europaea* L.	--	86–90	1,470
Palm	*Elaecis guineensis* L.	--	49–59	--
Palm Kernel	*Elaecis guineensis* L.	--	204–207	510

† From: Martin et al. (1976).
‡ Iodine number. Fatty acids combine with iodine in proportion to the number of double bonds present, i.e., the larger the iodine number the more unsaturated the fatty acid.

Industrial uses of vegetable oils encompass a diverse group of products such as pharmaceuticals, paints and other finishes, oil cloth, linoleum, plastics, and soaps. Glycerine, a component of vegetable oil, is used in the manufacture of explosives, foods, and cosmetics.

Vegetable Oil Quality

Oils and fats have similar structures, but those that melt below room temperature and are normally liquids are called oils. Oils and fats are made of triglycerides, i.e., a combination of glycerine and three fatty acids. A generalized formula for vegetable oil is shown in Fig. 1.11.

Fatty acids are long chain compounds of carbon, hydrogen, and oxygen. If the carbon atoms are bound to hydrogen or other carbon atoms, the fatty acid is saturated. If there are double bonds present (the carbon atoms are not bound to enough hydrogen) the fatty acid is unsaturated (Fig. 1.12). Fatty acids will combine with iodine in proportion to the number of double bonds present. The "iodine value" gives information about the degree of unsaturation of vegetable oils. For example, coconut oil contains mostly saturated fatty acids and has an iodine value of 8–12 while polyunsaturated corn oil has an iodine value of 115–130 (Table 1.13).

Oils that have a low iodine value are referred to as nondrying oils, i.e., they will not rapidly absorb oxygen from the air. Oils with a high iodine number are called drying oils because they are capable of ab-

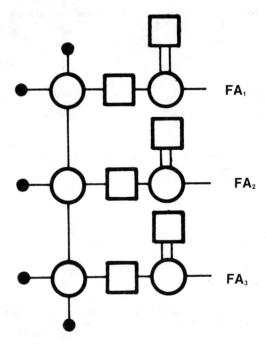

Fig. 1.11—Configuration of carbon (○), hydrogen (●), and oxygen (□) atoms and fatty
acid attachment sites on a typical triglyceride.

sorbing oxygen at their double bonds and will form an elastic film.
Drying oils such as tung (*Aleurites fordii* L.) and linseed have been
used extensively in paints and varnishes.

The degree of saturation in fats and oils has received increasing
attention by health specialists. Dietary fats high in saturated fatty
acids may lead to a cholesterol buildup in the body, which has been
associated with heart disease. Although this concept is still contro-
versial, it has increased the demand for polyunsaturated oils such as
vegetable oils over the more saturated animal fats.

Vegetable Oil Refining and Product Development

Oils are extracted from seeds by extrusion (mechanically squeez-
ing the oil out) or by solvent extraction (dissolving the oils in a suitable
organic solvent and distilling the solvent off). In some cases a combination of extrusion and solvent extraction may be used (Fig. 1.13).

The objective of the edible fats refiner is to produce oils and/or fats
that are pure and free of taste, smell, and color. After extraction, fat-
insoluble materials including gums, resins, and protein are removed by
a precipitation step called degumming. The oil is then neutralized to
remove any free fatty acids which could cause rancidity. To make the

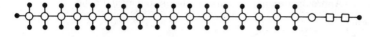

Stearic Acid, Saturated Fatty Acid

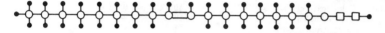

Oleic Acid, Monounsaturated Fatty Acid

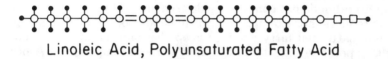

Linoleic Acid, Polyunsaturated Fatty Acid

Fig. 1.12—Configuration of carbon (○), hydrogen (●), and oxygen (□) atoms in fatty acids.

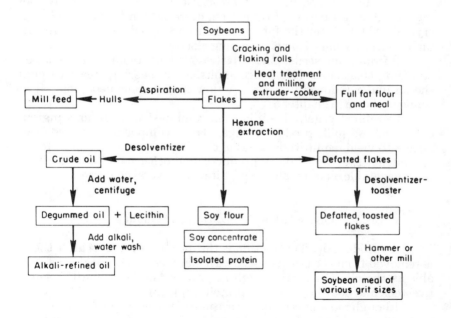

Fig. 1.13—Oil extraction flow diagram.

Table 1.14—Approximate composition of soybeans and meal products.†

Component	Protein	Fat	Carbohydrate	Ash
			%	
Whole Bran	40	21	34	4.0
Cotyledon	43	23	29	5.0
Hull	8	1	86	4.3
Meal	Min	Min	Max (Fiber)	
Cake (extruded)	41	3.5	7.0	
Flakes (extracted)	44	0.5	7.0	
Dehulled flakes (extracted)	49	0.5	3.5	
Mill feed (separated hulls)	13	--	32	
Mill run (separated hulls)	11	--	35	

† From: Cowan (1973).

oil colorless it must be bleached. Bleaching is accomplished by filter-
ing the oil through an adsorbent such as Fuller's earth (a hydrated
aluminum silicate).

Oils often do not contain the desired amount of saturation for certain products. For example, margarine and cooking fats must contain
enough saturated fatty acids to keep the material at a high enough
melting point so that it will not melt at room temperature. A process
called hydrogenation adds hydrogen to the unsaturated oils. Hydrogenation can bring an oil to a specific degree of saturation, thus allowing margarine producers to use oils from different seed sources that
may differ in level of saturation.

Oil is further modified by winterization. Oil is winterized by cooling the oil over a period of time. Cooling results in formation of large
crystals of fat. After the fat crystals are removed the oil can remain
under refrigeration without becoming cloudy.

Oil from some seeds, particularly soybeans, requires a special de-
odorizing treatment. This is accomplished by passing steam through
the oil under reduced pressure. A flow diagram for vegetable oil re-
fining is shown in chapter 5.

The refined vegetable oil is then combined with various ingredients, such as milk products, eggs, flavor compounds, and coloring
agents to yield a multitude of salad dressings, margarines of different
consistencies, artificial dairy products, and other food items. Bulk refined vegetable oil is a major export item of several countries.

Oilseed Meal Utilization and Quality

A byproduct of oil extraction is the meal that remains after the oil
is removed from the crushed seeds. In many cases the meal is as valuable as the oil. The meal is high in protein and is primarily used in
livestock and poultry feeds as a protein supplement.

Although most oilseed meals are used for feed, soybean meal is the
major one used in the USA and some foreign countries. Several types
of soybean meal containing different amounts of protein are used in
the livestock and poultry industry. Table 1.14 gives the composition of
soybeans and several soybean meals. Soybeans also contain several

anti-quality compounds, including a trypsin inhibitor, but these are normally destroyed when the meal is cooked with moist heat. Cottonseed meal also contains an anti-quality component called gossypol which is toxic to monogastric animals. Fortunately gossypol can be extracted from the meal, resulting in an excellent feed supplement.

Soybeans, peanuts, and other legumes contain more lysine but less methionine than the cereal grains so a mixture of corn and soybean meal, for example, would supply monogastric animals adequate amounts of essential amino acids that would normally be limiting in a corn diet. It is possible to improve the amino acid balance of human diets in developing nations by mixing cereal and legumes. Oilseed meals, because of their high protein content have contributed to the development of textured protein products or artificial meat. For example, by extruding soybean flakes through different types of dyes and adding flavor and coloring compounds, artificial bacon, ham, sausage, chicken, and beef can be produced (Fig. 1.14). Although the practicality of these products is yet to be determined, unflavored soybean protein has been used to extend ground beef when beef prices are high.

Peanut or Groundnut Utilization and Quality

In major peanut (groundnut) producing countries like India and Nigeria, peanuts are used for oil both for domestic use as well as export. In the USA peanuts are used mostly for food purposes because of a higher market value. The major uses of peanuts are peanut butter, confectionary nuts, salted nuts, and peanuts roasted in the shell. Most peanuts utilized for oil are either split, broken, or otherwise inferior nuts or surplus peanuts.

Peanut butter is a product of the USA. The origin and development of peanut butter are not well known, but apparently the commodity got its start in the late 1800's. Today about one-half of the USA's edible peanut crop goes into peanut butter. Georgia and Alabama grow nearly two-thirds of the peanuts used for peanut butter (Woodroff, 1973). Peanut butter is used in a multitude of ways, including sandwiches, baked goods, candy, and ice cream.

Quality requirements for peanut butter manufacturing are mostly physical. The peanuts must be shelled, plump, clean, free of contaminants and diseases, and preferably unbroken so that roasting will be uniform. Peanut butter is made by dry-roasting, blanching, and grinding the peanuts. Salt is usually added to enhance flavor. Hydrogenated fat and dextrose may be added to alter consistency. Corn syrup or glycerine is often added to keep the oil from separating, and lecithin or antioxidants are added to control rancidity (Woodroff, 1973).

Peanut butter taste and texture will vary with the amount of roasting and/or the flavor amendments that may be added. Some peanut butter manufacturers also include peanut particles to give a grainy or crunchy texture.

Fig. 1.14—Various textured vegetable protein simulated meats.

Salty peanuts are the second largest use of peanuts. These are plump, clean, shelled peanuts that have been roasted and salted. Salted peanuts are also utilized in various different ways. Over 1.3 billion kg of candy are produced in the USA annually and nearly 113 million kg of peanuts are used in candy production (Woodroof, 1973). Although there is controversy over the essentiality and wholesomeness of confectionaries, the addition of peanuts certainly improves the nutritive value of these products.

SUMMARY

The majority of the earth's 4 billion human inhabitants depend on crop plants for most of their dietary needs as well as the needs of some domesticated animals. Crop plants also provide a multitude of diverse industrial products for human consumption. Because of the diversity of crop plant utilization, many different quality parameters have been developed.

Quality criteria for crops have often been based on the physical requirements of food processors rather than on human or animal nutritional requirements. Consequently, crops have been developed to meet structural specifications for optimum milling but may result in a flour or meal deficient in lysine. Efforts are now being made to improve the nutritional status of major food crops.

Utilization of crop plants varies greatly throughout the world. While corn and sorghum are used mainly for animal feed in the USA, they are main staples for many people of the world. Often people in less developed countries suffer nutritional deficiencies because they are unaware of quality attributes of crops. A corn-based diet may be limiting in lysine, but addition of a legume rich in lysine can improve the diet.

Many industrial uses of crops have been made possible because of an improved understanding of quality attributes of cereals and oilseeds. Vegetable oil refiners have made great strides in produce development and quality manipulation.

Crop quality is a rapidly expanding science. A knowledge of quality will be important in developing an adequate nutritional status for an ever-increasing world population.

PHOTO CREDITS

Fig. 1.1 & 1.10—from Chrispeels and Sadava (1977), courtesy of W. H. Freeman and Co., San Francisco; Fig. 1.2, 1.3 & 1.6—courtesy of Wheat Flour Institute, Chicago; Fig. 1.7— from Reiners et al. (1973), courtesy of American Association of Cereal Chemists, St. Paul, MN; Fig. 1.13—from Cowan (1973); Fig. 1.14, courtesy of Miles Laboratories Chicago.

SUGGESTED READING

American Peanut Research and Education Association Inc. 1973. Peanuts— culture and uses; a symposium. Stillwater, OK.

Bauman, L. F., P. L. Crane, D. V. Glover, E. T. Mertz, and D. W. Thomas (ed.). 1975. High-quality protein maize. Dowden, Hutchinson, and Ross, Inc., Stroudsburg, PA.

Caldwell, B. E. (ed.). 1973. Soybeans: improvement, production, and uses. Agronomy 16. Am. Soc. Agron., Inc., Madison, WI.

Cereal Foods World. Am. Assoc. Cereal Chem., St. Paul, MN.

Houston, D. F. (ed.). 1972. Rice chemistry and technology. Am. Assoc. Cereal Chem., St. Paul, MN.

Hulse, J. H., and E. M. Laing. 1974. Nutritive value of triticale protein. Int. Devel. Res. Centre, Ottawa, Canada.

Inglett, G. E. (ed.). 1970. Corn: culture, processing products. The AVI Publ. Co. Inc., Westport, CT.

————. 1974. Wheat: production and utilization. The AVI Publ. Co. Inc., Westport, CT.

Jugenheimer, R. W. 1976. Corn improvement, seed production, and uses. John Wiley and Sons, Inc., NY.

Milner, Max (ed.). 1975. Nutritional improvement of food legumes by breeding. John Wiley and Sons, NY.

Pomeranz, Y. (ed.). 1971. Wheat chemistry and technology. Am. Assoc. Cereal Chem., St. Paul, MN.

————. 1973. Industrial uses of cereals; symposium proceedings. Am. Assoc. Cereal Chem., St. Paul, MN.

————. 1976. Advances in cereal science and technology, Am. Assoc. Cereal Chem., St. Paul, MN.

Simmonds, N. W. (ed.). 1976. Evolution of crop plants. Longman, NY.

Wall, J. S., and W. M. Ross (ed.). 1970. Sorghum production and utilization. The AVI Publ. Co. Inc., Westport, CT.

LITERATURE CITED

Adair, C. R. 1972. Production and utilization of rice. *In* D. F. Houston (ed.) Rice chemistry and technology. Am. Assoc. Cereal Chem., St. Paul, MN.

Anderson, R. A. 1970. Corn wet milling industry. *In* G. E. Inglett (ed.) Corn: Culture, processing, products. The AVI Publ. Co., Inc., Westport, CT.

————. 1974. Wet-processing of wheat flour. *In* G. A. Inglett (ed.) Wheat: production and utilization. The AVI Publ. Co., Inc., Westport, CT.

Brekke, O. L. 1970. Corn dry milling industry. *In* G. E. Inglett (ed.) Corn: culture, processing, products. The AVI Publ. Co., Inc., Westport, CT.

Brockington, S. F. 1970. Corn dry milled products. *In* G. E. Inglett, (ed.) Corn: culture, processing, products. The AVI Publ. Co., Inc., Westport, CT.

Chrispeels, M. J., and D. Sadava. 1977. Plants, food, and people. W. H. Freeman and Company, San Francisco, CA.

Cotton, L. (ed.). 1970. Old Mr. Boston De Luxe Official Bartender's Guide. Mr. Boston Dist. Corp. Publ., Boston, MA.

Cotton, R. H., and J. G. Ponte, Jr. 1973. Baking industry. *In* G. E. Inglett (ed.) Wheat: production and utilization. The AVI Publ. Co., Inc., Westport, CT.

Cowan, J. C. 1973. Processing and products. *In* B. E. Caldwell (ed.) Soybeans: improvement, production, and uses. Agronomy 16. Am. Soc. Agron. Inc., Madison, WI.

Gariboldi, F. 1973. Parboiled rice. *In* D. F. Houston (ed.) Rice chemistry and technology. Am. Assoc. Cereal Chem., St. Paul, MN.

Hahn, R. R. 1970. Dry milling end products of grain sorghum. *In* J. S. Wall and W. M. Ross (eds.) Sorghum production and utilization. The AVI Publ. Co., Inc., Westport, CT.

Heiser, C. B., Jr. 1976. Sunflowers. *In* N. W. Simmonds (ed.) Evolution of crop plants. Longman, NY.

Martin, J. H., W. H. Leonard, and D. L. Stamp. 1976. Principles of field crop production. 3rd ed. Macmillan Publ. Co., NY.

Matsuo, R. R. 1975. Uniqueness of pasta. Cereal Foods World 20:485.

Medcalf, D. G. 1973. Structure and composition of cereal components as related to their potential industrial utilization. *In* Y. Pomeranz (ed.) Chemical and industrial uses of cereals. Am. Assoc. Cereal Chem., St. Paul, MN.

National Academy of Sciences. 1964. Feed composition. Joint United States-Canadian tables. Publ. 1232. NAS-NRC, Washington, DC.

Peterson, G. A., and A. E. Foster. 1973. Malting barley in the United States. Adv. Agron. Academic Press, NY.

Pomeranz, Y. 1972. Rice in brewing. *In* D. F. Houston (ed.) Rice chemistry and technology. Am. Assoc. Cereal Chem., St. Paul, MN.

————. 1973. From wheat to bread: a biochemical study. Am. Sci. 61:683–692.

Reiners, R. A., J. S. Wall, and G. E. Inglett. 1973. Corn proteins: potential for their industrial use. *In* Y. Pomeranz (chm.) Industrial uses of cereals. Am. Assoc. Cereal Chem., st. Paul, MN.

Scrimshaw, N. S., and V. R. Young. 1976. The requirements of human nutrition. Sci. Am. 235:51–64.

Walsh, D. E., and K. A. Gilles. 1973. Macaroni products. *In* G. E. Inglett (ed.) Wheat: production and utilization. The AVI Publ. Co., Inc., Westport, CT.

Woodroof, J. G. 1973. Peanuts: production, processing, products. The AVI Publ. Co., Inc., Westport, CT.

World Health Organization. 1973. Energy and protein requirements. Tech. Rep. Ser. 522. WHO, Geneva.

Chapter 2

Fiber Crops

H. H. RAMEY, JR.

SEA-USDA and University of Tennessee
Knoxville, Tennessee

Clothing and textiles are second only to food as absolute necessities for human existence. Plant fibers have a long history of use in clothing and textiles. Cotton is the plant fiber most commonly used in clothing and is also used extensively in household textiles. Some of the bast fibers are used in clothing and household textiles. Other bast fibers and leaf fibers are used for packaging.

Cotton lint is a fiber which grows from the seed surface in the pods, or bolls, of a bushy mallow plant. Four species of the genus *Gossypium* produce the cotton lint of commerce. Two of the species, *G. arboreum* L. and *G. herbaceum* L., are natives of the Old World; both are grown on the western side of the Indian subcontinent. *Gossypium arboreum* is cultivated eastward in Asia to Manchuria, and *G. herbaceum* is grown westward through Asia Minor. The other two species are indigenous to the New World. *Gossypium barbadense* L. is native to South America and is now grown primarily in Peru, Egypt, and the USA. *Gossypium hirsutum* L. is a native of Mexico and Central America and is now grown worldwide in tropical, subtropical, and temperate regions. Fiber from *G. hirsutum* constitutes the majority of the lint produced in the world.

Bast fibers are obtained from the inner bast tissue or "bark" of the stems of herbaceous, dicotyledonous plants and are often designated as "soft" fibers. Flax, *Linum usitatissimum* L. and hemp, *Cannabis sativa* L., are the more important bast fibers used for textiles. Production of flax for fiber is limited primarily to Europe, the USSR being the major producer. Hemp production is concentrated in Europe, but it can also be found in Turkey, Korea, Taiwan, Japan, and Chile. The USSR is the major producer of hemp.

Jute, *Corchorus* spp., and kenaf, *Hibiscus cannabinus* L., are also bast fibers, but they are largely used for packaging material. Kenaf is also extensively used for making paper. Bangladesh and India are by far the major producers of jute fiber, whereas Thailand, Brazil, China, India, and the USSR are the principal kenaf-producing countries.

Leaf fibers are obtained from the leaves and leaf sheaths of mono-cotyledonous plants and are often called "hard" fibers. Abaca or Manila hemp, *Musa textilis* Nee; sisal, *Agave sisalana* Perr.; and hene-quen, *Agave fourcroydes* Lem., are the principal hard fibers and are used primarily for rope and cordage. Abaca is produced commercially almost entirely in the Philippines, although it is also grown in several of the wetter tropical countries. Sisal, though originally a native of Mexico, is grown mostly in Africa with some production in Brazil, Haiti, Taiwan, Indonesia, and Venezuela. Henequen is native to Mexico, and there its major commercial production is centered today.

COTTON

History

Cotton has been used as a textile material for more than 5,000 years. Fragments of cotton bolls, found in Mexico, that were dated from 5800 B.C. resembled modern types. However, there is some question as to whether cotton was used for textile purposes that early. Textile fragments made from cotton fibers have been found by archaeologists that date from 3500 B.C. in Mexico, 3000 B.C. in the Indus Valley, 2500 B.C. in Peru, and 500 B.C. in the southwestern United States. Weaving of cotton in Egypt was mentioned by Isaiah about 750 B.C. The Medo-Persian Empire used cotton in the seventh century B.C., particularly in Susa. Alexander the Great returned from India in 327 B.C. with robes of cotton. Awnings of cotton were used in Roman theaters by 60 B.C.

Cotton was used as a decorative plant in China by 700 A.D. Marco Polo in his great trip, 1275 to 1297, found cotton production and tex-tiles from it in Kurdistan, western China, and India. The Moors brought cotton culture to Spain in 712. Cotton was introduced into other parts of Europe by the returning Crusaders.

Columbus found cotton being grown in the Caribbean during his first trip of discovery in 1492. A sizable cotton industry was found by Cortez during the invasion of Mexico in 1519.

Cotton was first planted in 1607 at Jamestown, VA. Continuous culture of cotton in the eastern USA dates from 1621. Production of cotton was primarily for home use until Eli Whitney invented the cotton gin in 1793. Overcoming the difficulty of separating the cotton fiber from the seed provided the impetus for greatly expanded cotton production in the years immediately following the invention of the cotton gin. Production of cotton in other parts of the world was stimu-lated by cotton shortages caused by the War Between the States in 1861 to 1865.

Structure

A cotton fiber originates as a tubular outgrowth of a single cell on the seed coat. The cell begins development the day of flowering (Fig. 2.1), and elongation of the cell proceeds for some 15 to 25 days (Fig.

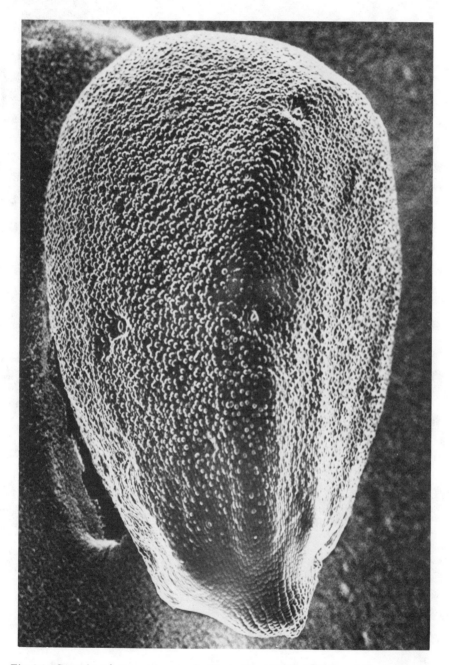

Fig. 2.1—Scanning electron micrographs illustrating (A) The initiation of fibers on most of the seed surface on the day of anthesis (×50)

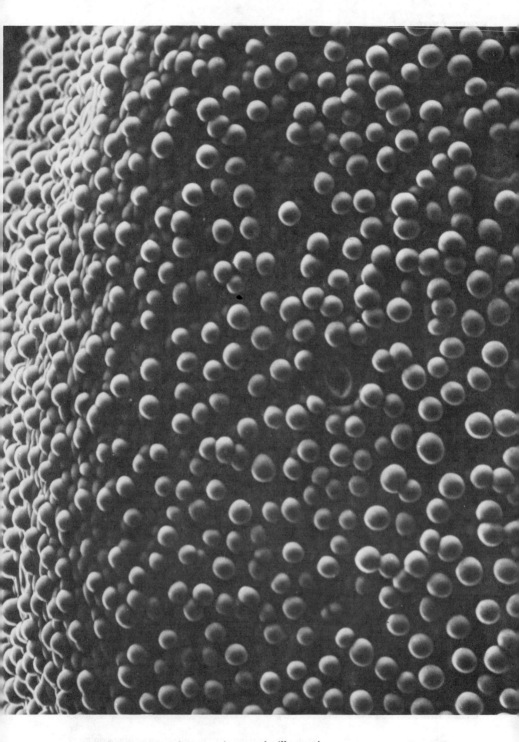

Fig. 2.1—Scanning electron micrographs illustrating
(B) The spheroid-like appearance of the fibers on the day of anthesis (×500)

and (C) The tube-like growth of the fibers at   3 days after anthesis ($\times 500$).

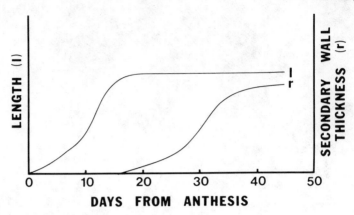

Fig. 2.2—Cotton fiber development. The individual fiber cell elongates for some 20 days after anthesis. As the fiber approaches its ultimate length (*l*), secondary wall formation begins and continues for 20 or more additional days (r).

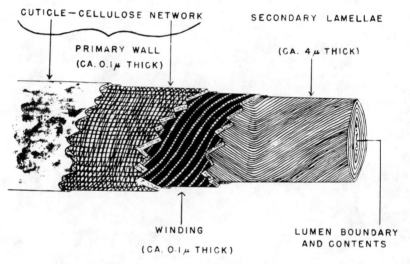

Fig. 2.3—Schematic diagram of the cotton fiber showing its structural features.

2.2). The ultimate length of the cell may exceed 3,000 times its diameter. During the elongation phase, the structural component of the fiber cell is the primary wall. As elongation ceases, secondary wall development begins and continues for 20 or more additional days. The secondary wall constitutes about 95% of the cellulose of the mature fiber and is the component of major importance in the performance of the fiber in use.

The secondary wall is almost pure cellulose, made up of long molecules of polyglucose. Layers of cellulose are added daily to the interior of the cell wall (Fig. 2.3). Diurnal temperature fluctuations cause changes in density of the cellulose laid down in each layer.

The cellulose molecules contain up to 3,000 glucose units. When laid down in the secondary wall, the molecules aggregate into micro-

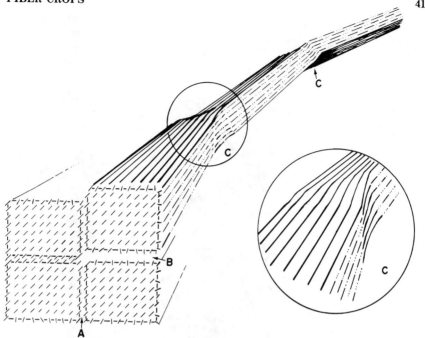

Fig. 2.4—Schematic representation of a microfibril to illustrate (A) Coalesced surfaces of high order, (B) Readily accessible slightly disordered surfaces, and (C) Readily accessible surfaces of strain distorted tilt and twist regions. Individual units in cross-section contain from 8×10 to 12×14 cellulose molecules.

fibrils which are highly organized rectangular structures, containing from 8×10 to 12×14 molecules in cross section and are highly crystalline (Fig. 2.4). Hydrogen bonding occurs among the microfibrils, leading to a structure of higher order. Some cellulose molecules extend from one microfibril to another. In areas between microfibrils, the cellulose molecules are not organized into rectangular structures. These amorphous regions, however, provide the links between the highly ordered portions of the cell wall.

The elementary cellulosic units are laid down in a "leno-weave" pattern in the primary wall (Fig. 2.3). This pattern resembles the interweaving in burlap bagging. The first layer of the secondary wall is called the winding layer. In portions of this layer, the leno-weave persists; in other portions, the microfibrils are laid down in a spiral manner around the fiber. The spiral is at an angle of 55° to the long axis. As further layers of the secondary wall are laid down, the microfibrils are all oriented at an angle to the major axis of the fiber. The angle of orientation decreases from the 55° in the winding layer to 36° in the middle to 18° in the interior of the wall. Thus, the secondary wall of the cotton fiber is an assemblage of crystalline elementary microfibrils that are wrapped in helical layers.

At random intervals, the spiral structure reverses direction (Fig. 2.5). The angle to the long axis of the fiber is the same, whether right or left, within a layer. The reversals occur on an average of 20 to 30 per cm of fiber length.

The perimeter of the developing cotton fiber is from 60 to 90 μm depending upon the genetic type. As the mature fiber dries, the primary wall shrinks and the whole structure collapses. The protoplasm dries into a tiny proteinaceous remnant in the lumen. The perimeter of

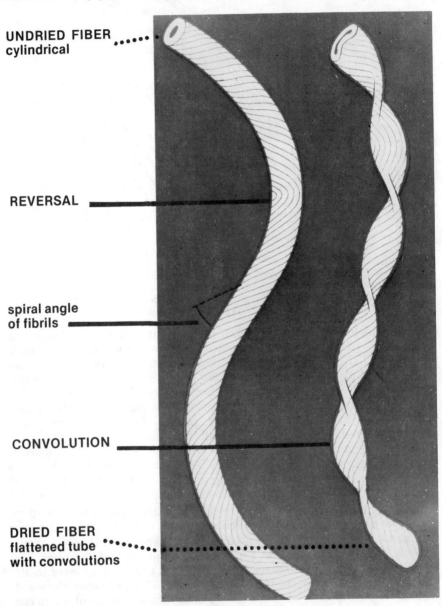

UNDRIED FIBER
cylindrical

REVERSAL

spiral angle
of fibrils

CONVOLUTION

DRIED FIBER
flattened tube
with convolutions

Fig. 2.5—The cotton fiber is originally cylindrical but collapses into a flattened tube upon drying. Convolutions occur in the dried fiber, and they change direction at least at every second reversal point.

the dried fiber is from 40 to 60 μm, again depending upon genetic type. The cross-section of the collapsed dried fiber varies from bean or kidney-shaped to oval or nearly circular, depending upon the wall thickness (Fig. 2.6).

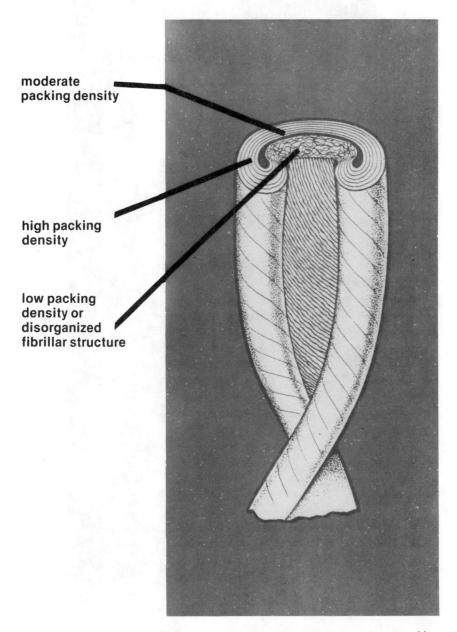

moderate
packing density

high packing
density

low packing
density or
disorganized
fibrillar structure

Fig. 2.6—Morphological model of cotton fiber showing areas of high, moderate, and low packing density.

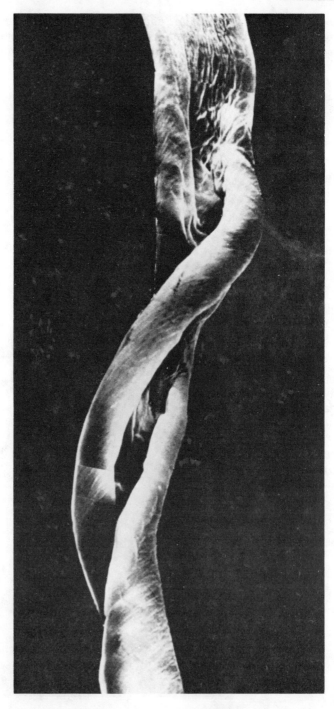

Fig. 2.7—Portion of a collapsed, convoluted fiber.

The never-dried fiber is a highly plasticized, relatively strain-free structure. The elementary structural units are surrounded by water, and this provides considerable plasticity. Upon initial drying, some irreversible hydrogen bonding occurs among the microfibrils and greatly reduces the pliability of the structure. However, some of the hydrogen bonding is reversible; thus, some plasticity is retained in the dried fiber. However, the plasticity of the undried fiber is much greater than that of a wet, though once-dried, fiber.

As the cell wall collapses upon initial drying, the cotton fiber becomes twisted and convoluted (Fig. 2.5, 2.7). The presence and number of convolutions depend upon wall thickness and number of reversals. There are few or no convolutions in thin-walled fibers, numerous convolutions in fibers of intermediate wall thickness, and few in thick-walled fibers. Convolutions change directions at each reversal point or every second reversal point.

This collapse of the cell wall into a smaller structure sets up stresses within the fiber. In the kidney shape, the microfibrils are more densely packed in the ends, and less densely packed on the concave side. The less densely packed portions of the secondary wall are more readily attacked by enzymes and degrading substances (Fig. 2.6).

A cotton fiber is primarily cellulose but contains other organic materials as well. A typical mature cotton fiber is composed of the following:

Constituent	Dry weight (%)
Cellulose	94.0
Protein	1.3
Pectic substances	1.2
Ash	1.2
Wax	0.6
Sugars	0.3
Others, including organic acids	1.4

The outer surface of the mature fiber is covered with a cuticle layer composed of waxes. In textile processing, fibers move past one another and the waxy cuticle acts as a lubricant.

Fiber Quality

Cotton lint is a biological product and as such is variable. Differences in quality were noted as early as in the 13th century in India.

As the Industrial Revolution progressed in England and the rest of Europe, cotton was imported from various areas of the world. The locality of growth was initially used to designate quality. In the USA, such terms as "Texas Blacklands," "Benders," "Rivers," "Peelers," "North Georgias," and "Canebreakers" were used as quality descriptions. However, these designations soon proved insufficient to provide for orderly marketing.

The U.S. Cotton Standards Act of 4 March 1923, authorized the Secretary of Agriculture to establish standards for the classification of cotton by which its quality could be judged or determined for commercial purposes. It also made it unlawful to indicate for any cotton a grade or other class by a name description or designation not used in those standards.

The Universal Standards Agreement entered into by the U.S. Department of Agriculture and 14 cotton associations and exchanges in Europe, Japan, and India provided for the adoption, use, and observance of the Universal Standards in the classification of U.S. upland cotton. The Universal Standards are the Official Cotton Standards of the USA, and those standards are used for a major portion of international commerce.

Factors of grade

Grade is a function of three factors—color, leaf or foreign matter, and preparation.

Color. When an upland cotton boll opens, its fiber is white or slightly creamy in color. However, the cotton plant itself is indeterminate; bolls mature over a time span of 4 or more weeks. In harvesting with spindle pickers, at least three-fourths of the bolls are allowed to open before harvest begins. When mechanical strippers are used, harvest is delayed until all the bolls are open. Continued exposure of seed cotton to weather conditions in opened bolls in the field or the action of microorganisms causes the fiber to lose its brightness gradually and become dull and darker in color.

Under extreme conditions of weathering and microorganism attack, the color may become a dark bluish gray.

When upland cotton fiber development is stopped prematurely by drought, frost, or other causes, a yellow color, which varies in depth, may develop. The action of insects and microorganisms or soil depositions from rain may cause the fiber to become discolored or spotted. Oil or grease from harvesting machinery and the exudates from plant parts crushed during harvest may cause discolorations.

Regardless of cause, any departure from the bright color of normally opened cotton bolls indicates a deterioration in quality. Dull or darker-colored cotton fiber receives a lower grade.

Leaf or Foreign Matter. Leaf includes dried and broken plant parts of various kinds which became entangled with the cotton fibers during harvest. Parts of leaves and bracts are predominant, but other materials, such as stems, bark, parts of burs (carpels), whole seed, parts of seed, motes (immature or undeveloped seed), grass, sand, and dust are sometimes found. Large particles are generally less objectionable as they are more easily removed by cleaning processes during ginning and yarn and fabric manufacture. Small particles, called "pin" or "pepper" trash, are exceedingly difficult to remove.

Table 2.1—Codes for the grades of upland cotton.†

| | | | Color group | | | | |
Name	Plus	White	Light Spotted	Spotted	Tinged	Light Gray	Gray
Good Middling		11	12	13		15	16
Strict Middling		21	22	23	24	25	26
Middling	30	31	32	33	34	35	36
Strict Low Middling	40	41	42	43	44	45	46
Low Middling	50	51	52	53	54		
Strict Good Ordinary	60	61					
Good Ordinary	70	71					

† Adapted from summary of cotton fiber and processing test results, crop of 1977. USDA Agric. Marketing Service, Cotton Division, Memphis, TN.

The amount of foreign matter remaining in lint cotton after ginning (the process of removing fibers from the seed) represents a loss since it must be removed in the manufacturing process. The particles which are not removed in the manufacturing process detract from the appearance of the yarn or fabric offered for sale. Lint having the smallest amount of trash, other properties being equal, has the highest value for manufacture of fabrics.

Preparation. "Preparation" is the term used to describe degree of smoothness or roughness with which cotton is harvested, ginned, or both. Usually smoothly harvested and/or ginned cotton results in less waste and produces more uniform yarn than roughly harvested or ginned cotton. Improperly adjusted machinery and high moisture content during processing may cause rough preparation.

Grades provide an indication of the non-lint content of a sample of cotton and reflect the waste that may occur during processing. Thirty-four grades of upland cotton are recognized. There are seven names and seven color groups. The official designation is the name followed by the color group, such as Strict Middling Light Spotted, Strict Low Middling Light Gray or Strict Good Ordinary Plus. A code is frequently used to represent the grade. The codes for the recognized grades are listed in Table 2.1. There are nine grades of American Pima cotton and these are designated one through nine as American Pima No. 1 through American Pima No. 9. Samples of the white, spotted, and tinged upland cotton grade standards and American Pima No. 2 through American Pima No. 9 are available from Agricultural Marketing Service, U.S. Department of Agriculture. The other grades are described in reference to the samples of the grade standards.

The bale is the marketing unit for U.S. cotton. Thus, each bale must be graded. In most other producing countries, the marketing unit is from 100 to 300 bales. Bales vary from 180 to 390 kg, depending on the practice or custom in the producing country; the nominal U.S. bale is 220 kg. When the marketing unit is large, a more elaborate grading system is feasible. The USSR uses fiber strength, maturity, and moisture content in grading. Grades used in Brazil include length as part of the grade designation.

Staple

Staple length of cotton is defined in the original order promulgating staple standards as:

"The length of staple of any cotton shall be the normal length by measurement, without regard to quality or value, of a typical portion of its fibers under a relative humidity of the atmosphere of 65% and a temperature of 70°F (21°C)."

Staple length is an important quality because long fibers are required to produce fine yarns. Experience is essential to determine the normal length because any sample will have in it fibers of many different lengths. For a given length of the longer fibers, those samples that have a lower proportion of short fibers will be designated longer than those that have a higher proportion of short fibers. Thus, the staple length is an index of quality.

The linear inch is the unit used to designate the length of fibers in a sample. Staple lengths are reported in thirty-seconds of an inch for upland cotton. The range is from 13/16 to 1 1/4 in. American Pima cotton is reported in sixteenths of an inch and ranges from 1 3/8 to 1 1/2 in. There are official Cotton Standards for each of these staple lengths and samples of the standards are available from Agricultural Marketing Service, U.S. Department of Agriculture.

Because of international agreements, the inch is used to designate staple length in international trade. However, some countries such as USSR use mm for length. The basis for the length of USSR cotton is different than that of USA cotton in which the length is determined by reference to the official standards. For example, USSR cotton designated 31/32 mm would be designated 1 3/32 in by upland cotton standards. Converting the inch to mm would result in 27.8 mm for the cotton having a designation of 31/32 mm. Staple length should be designated in inches only in order to avoid confusion.

Instrument Measures

Although cotton is marketed on the basis of classer's subjective determination of grade and staple, instrumental measurements have been developed to supplement the hand and eye judgment of the classers.

Fineness and maturity. The weight per unit length of a fiber is an important quality and variations in fiber perimeter or in wall thickness can alter weight per unit length. Within the upland cottons, fiber perimeter is relatively constant. Therefore, wall thickness or maturity is the major contributor to changes in weight per unit length. Processing waste of immature cotton is higher during yarn manufacture. Immature cottons also cause trouble in finishing of fabrics, particularly in uniformity of appearance after dyeing.

Air-flow instruments are used to measure fineness and maturity. The commonly used instruments, Micronaire and Fibronaire, provide a rapid measure of fineness and maturity in combination; results from both instruments are reported in standard micronaire units. A standard weight of representative fibers is placed in the specimen holder of the instrument and compressed to a standard volume. The flow of air through the specimen is read on a scale calibrated in micronaire units. A fine or immature fiber has more resistance to air flow and a low micronaire reading will result. A coarse or mature fiber has less resistance to air flow, and the micronaire reading will be higher.

The micronaire scale is calibrated from 2.4 to 8.0. Readings within the range of 3.5 to 4.9 are considered optimal and this is referred to as the "premium range." Fibers having readings of 3.4 or less may cause processing problems. Fibers having readings of 5.0 or more may be too coarse for many applications, particularly for use in fine yarns. Micronaire readings are used in the marketing of the U.S. cotton in conjunction with grade and staple determinations. Cottons that have either low or high micronaire readings may be discounted.

Fiber Length. Although cotton fiber length is determined by classers using visual observation and comparison to official standard samples for marketing purposes, instruments which measure length are widely used in research and as an adjunct to a classer's staple length. The most commonly used length-measuring instrument is the Fibrograph. The specimen to be measured is combed out straight in beard form on special combs. The comb and cotton specimen are placed in the machine over a long narrow slot. A photoelectric cell is behind the slot, and light is directed on the specimen. The specimen is placed so that the slot is near the comb. The short and long fibers intercept the light so that only a small amount of light reaches the photoelectric cell. The specimen is slowly moved perpendicular to the slot so that fewer and fewer of the fibers intercept light. The movement of the specimen to the point at which 50% of the light is intercepted is noted. The length from the comb of this point on the specimen is the 50% span length. The length at which 2.5% of the light is intercepted is the 2.5% span length. The designation, span length, is used to indicate that this percentage of the fibers are the stated length or longer. The 2.5% span length approximates classer's staple length. However, for two samples that have the same 2.5% span length, the staple length of the one with the longer 50% span length may be longer than the one with the shorter 50% span length.

An additional measure of fiber length is the Uniformity Index. This is the ratio of 50% span length to 2.5% span length times 100. A Uniformity Index below 45 indicates that many lengths of fibers are in the specimen, whereas Uniformity Index above 50 is desirable and suggests that the fibers in the specimen are more nearly the same length.

Fiber Strength. Fiber strength cannot be determined accurately by the classer. Instruments for determining strength of fiber bundles are frequently used in the textile industry. In making strength tests, cotton fibers are combed to parallelize them. A flat ribbon of fibers about 3 mm wide is formed and placed in a set of breaking clamps. The fibers are cut to a definite length, i.e., that which spans the clamps. The specimen is broken and the force required to do so is recorded. The broken ribbon is weighed, and the mass is recorded. The breaking force is divided by the specimen mass to calculate tenacity, which is expressed in millinewtons (mN) per tex (tex is the weight in grams of 1,000 m of fiber). Fiber strength may also be reported in pounds per square inch (psi). Since the cross-sectional area is not measured in strength determinations, the psi is a converted measure based upon an empirical relationship.

In the strength determination, the jaws of the breaking clamp may be placed together (zero gauge), or a spacer of a finite width, usually 1/8 inch, or 3.2 mm, can be placed between the jaws of the breaking clamp (1/8 gauge). The psi measure is calculated only when the strength determination is made at zero gauge. Fiber tenacity is reported for either zero or 1/8 gauge breaks.

Strength of cotton fibers is approximately 50% less at 1/8 gauge than at zero gauge. Several factors may contribute to this loss of strength with increased distance between the jaws of the breaking clamps. A major factor is the reversals. When linear stress is applied, cotton fibers rupture near a reversal point (Fig. 2.8). With 1/8 gauge breaks, up to 10 reversals in a fiber can occur between the jaws. Thus, more weak points are subject to rupture. A reduction in the force required for break is not unexpected. The typical range for zero gauge fiber tenacity is 300 to 500 mN/tex and 1/8 gauge, 150 to 300 mN/tex.

With other fiber properties the same, the cotton that has a higher fiber tenacity will produce a yarn that has a higher tenacity. This is particularly important in producing yarns on open-end spinning equipment because yarns produced on these machines tend to have lower tenacity than yarns produced on ring spinning frames.

Fiber Types

The Asiatic cotton fibers are short and usually coarse. The Egyptian cotton fibers are long, fine, and strong. Upland cotton fibers are intermediate in length, but vary considerably in coarseness and strength.

Asiatic Cotton

For the most part, cotton fibers produced by G. arboreum and G. herbaceum are shorter than an inch (25 mm) and have Micronaire readings of 3.0 to 7.0. Their major use in the USA is in surgical supplies.

Egyptian Cotton

The fiber produced commercially from G. barbadense is long, fine, and strong. The major area of production in the world is the Nile Valley of Egypt and the Sudan. About 1% of the USA crop (American

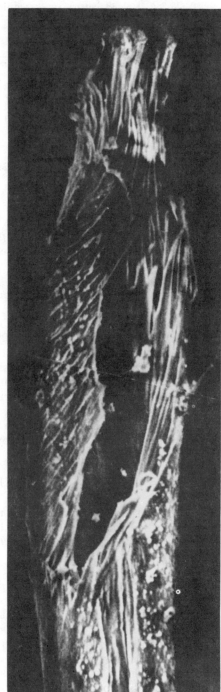

Fig. 2.8—Portions of two broken fibers.

Table 2.2—Upland cotton types within the U.S. crop and usual range of fiber properties.†

Type	Production area	Nominal staple length	Micronaire reading	Fiber tenacity
		in		mN/tex
1	Oklahoma and central and west Texas	15/16 to 1	3.2 to 4.6	170 to 190
1a	High Plains of Texas	15/16 to 1	3.2 to 4.0	190 to 210
1b	West Texas	31/32 to 1 1/16	3.2 to 4.8	180 to 210
2	Imperial Valley of California, Arizona, Lower Rio Grande Valley of Texas, and Mid-south and Southeastern States	1 1/32 to 1 1/8	3.8 to 4.8	180 to 200
2a	Mid-south and Southeastern States	1 3/32 to 1 5/32	3.8 to 4.8	190 to 210
2b	San Joaquin Valley of California	1 3/32 to 1 5/32	3.8 to 4.5	210 to 230
3	Far West Texas, New Mexico, and Arizona	1 1/8 to 1 3/16	3.2 to 4.0	220 to 240

† Adapted from Ramey (1974).

Pima) is of this type. The major use is in sewing thread, although a small amount is used in fabrics where silky smoothness, softness, and luster are desired. The staple length is 1 3/8 in and longer, Micronaire reading 2.8 to 4.0 and 1/8 gauge tenacity 280 to 300 mN/tex.

Upland Cotton

A major portion of the world's cotton production is upland cotton, *G. hirsutum.* In the USA about 99% of the production is from upland cultivars. Three types long recognized in the crop are differentiated by staple length. These are short (less than 1 in staple), medium (1 to 1 1/8 in), and long staple upland cottons (1 1/8 in and longer). Subtypes within the short and medium categories can be differentiated on the basis of fiber strength (Table 2.2).

The shorter staple types can be used only for coarser yarns. Fine yarns for sewing thread require long, fine fibers.

Uses

More than 90 specific textile products are made from cotton fibers alone or from cotton blended with other fibers. These are classed into three major categories, i.e., apparel, household, and industrial uses.

Apparel

Substantial quantities of cotton fibers are used for clothing. Some fabrics are made from all cotton such as denim for slacks or jeans and as handkerchiefs. However, many fabrics for apparel uses are made from cotton blended with man-made fibers, mainly the polyesters. Shirts, blouses, slacks, and dresses are examples of clothing made from blends. Cotton content will vary from 20 to 60%. Lightweight fabrics require longer fibers, but denim and corduroy can be made from the shorter fibers.

Household

Most towels and washcloths are made from 100% cotton. Sheets, pillowcases, bedspreads, tablecloths, and napkins may be made from cotton or cotton blended with man-made fibers. Fabrics made from blends are used in curtains, drapery, upholstery, and slip covers. Nylon, polyester, and rayon may be used as the other fiber in blends with cotton for these uses. The cotton content will vary from 20 to 85%.

Industrial

All-cotton fabrics are used for backing abrasive materials, canvas shoes, some awnings, tarpaulins, linings for sleeping bags, and for wiping and polishing cloths. Fabrics made from blends are used in automobile interiors, upholstery and vinyl hardtop backing, boat covers, luggage, and coated wall covering fabric. Cotton is also used for machinery belting reinforcement and for twine.

Cotton in Competition

Cotton is one of many fibers available for use in textile applications. The choice of a fiber for a specific use depends upon the properties of the fiber and its relative cost. Although prices for fibers fluctuate widely, cotton is frequently not the cheapest of the component fibers within a product.

Cotton has specific properties that make it the preferred fiber for many applications. In apparel, cotton increases the comfort of the garment. Fabrics containing high levels of polyester or nylon will develop static charges on cool, dry days while cotton is free from static electricity. Cotton takes up moisture rapidly and has a high saturation point compared to many man-made fabrics; it also gives up moisture readily. This latter property makes cotton fabrics feel cool. The water absorbency of cotton makes it ideally suitable for towels; rayon, for example, will take up more moisture than cotton (13 vs. 9%) but will not give it up as readily. Therefore, rayon towels will not dry as well as cotton towels and feel clammy to the touch after being used.

Cotton is readily destroyed by acids, but is highly resistant to bases. The strength of cotton is actually greater when wet than dry.

A smooth appearance is desired in many apparel uses. Untreated cotton wrinkles readily. For apparel usage, cotton fabrics are normally given a treatment to reduce its propensity for wrinkling. However, the resin treatment, called easy-care or durable press finish, reduces the strength of the cotton fibers by about 60% and reduces their resistance to abrasion. Man-made fibers are blended with cotton fibers to improve the strength and abrasion resistance of fabrics for apparel uses. When polyester is the other fiber in a blend with cotton, the best balance of wear, comfort, and appearance is achieved with about 60% cotton. Resins which impart a durable press finish react with the cellulose of cotton. Thus, the finish is only on the cotton part of the blended fabric.

When a cotton fabric does become wrinkled, it can be readily pressed to improve its appearance. In pressing a blend fabric with a hot iron, the surface of the thermoplastic fiber (nylon or polyester) will melt and become tacky. Thus, a blend fabric must be pressed at fairly cool temperatures.

Energy

All fibers require energy for their production and manufacture into fabrics. Cotton uses the sun's energy on the earth's surface to make the cellulose in its fiber. The cellulosic man-made fibers, rayon and triacetate, are made from certain woods. Energy for making their cellulose is also obtained from the sun. However, stored energy (as contrasted to energy currently falling on the earth's surface) is required to make textile products from the non-cellulosic, man-made fibers. Stored energy in the form of petrochemicals is used to make fibers such as polyester and nylon. The supply of petrochemicals is finite and decreasing.

The amounts of energy required to produce woven fabrics from cotton, rayon, and polyester are shown in Table 2.3. The energy required to produce polyester fabric is approximately 1.6 times as great as for cotton. Rayon also requires greater energy consumption compared to cotton.

Fashion

The relative use of textile fibers, particularly in apparel, is dependent upon fashion. Denims and corduroy, for example, are best made from cotton. When apparel made from such fabrics are in vogue, cotton will be used in greater quantities than when other fabrics are the fashion.

BAST FIBERS

History

Neolithic man used flax for making fishing nets, traps, bags, hammocks, and linen cloth. The ancient Egyptians were also users of linen. The original home of hemp was probably central Asia. Hemp has been grown for textile purposes in China for at least 4,500 years. Jute has been a widely used packaging fiber for only about 150 years; however, it has been grown for market in India for over 1,000 years. Use of kenaf

Table 2.3—Total energy consumption from raw materials to finished broadwoven fabric in kilowatt hours per kilogram of fiber (kilowatt hours per pound of fiber).†

Location of consumption	Cotton	Rayon (cellulosic)	Polyester (non-cellulosic)
Raw materials	3.09 (1.40)	2.93 (1.33)	19.25 (8.28)
Fiber production	11.42 (5.18)	35.71 (16.20)	16.00 (7.26)
Weaving mills (including spinning)	13.89 (6.30)	15.50 (7.03)	15.50 (7.03)
Finishing mills	15.39 (6.89)	18.78 (8.52)	18.78 (8.52)
Cumulative total	43.79 (19.86)	72.92 (33.08)	68.53 (31.09)

† All energy consumed is expressed in kilowatt hour equivalents to facilitate comparisons. Energy may be derived from natural gas, petroleum, coal, nuclear, or water power. Adapted from Gatewood (1973).

as a packaging fiber is more recent. Kenaf, used today primarily as a substitute for jute, may have been used as early as 4000 B.C. in some parts of ancient Africa and Asia.

Structure

Bast fibers are the fibers derived from the phloem of dicotyledonous, herbaceous plants. The individual cells elongate within the phloem tissue and subsequently develop secondary walls. The structure of an individual fiber cell resembles the cotton fiber. However, reversals are not present in the secondary wall of the bast fibers. Upon drying, the fiber cell does not collapse as it does in cotton. The lengths of bast fibers are at least 100 times their diameter. Their dimensions are as follows:

Fiber	Length, mm	Width, μm
Flax	16–30	7–20
Hemp	15–25	14–30
Jute	2–3	10–20
Kenaf	1.5–3	18–20

The major constituent of bast fibers is cellulose, but other compounds are present as well. Typical components expressed as percentages are as follows:

Fiber	Cellulose	Hemi-cellulose	Lignin
Flax	71	18	2
Hemp	74	18	4
Jute	67	18	13
Kenaf	74	--	11

Fiber Quality

Assessment of flax quality is done manually and visually. The main characteristics taken into account are fineness, softness, strength, density, color, uniformity, silkiness or oiliness, length, handle, and cleanliness. The quality of hemp is judged mainly by color and luster, but softness may also be considered. Species, strength, coarseness, color, length, ends, moisture content, and freedom from extraneous matter are factors in grading jute. The strength is partially estimated from the gloss, texture, and luster of the fiber. Color, length, fineness, strength, and cleanliness are considered in grading kenaf. The best kenaf will be equivalent to the medium grades of jute because the kenaf fiber is less flexible and more coarse.

All bast fibers must be removed from other stem tissue. This process, called retting, is accomplished by exposing the stems to moisture. Action by water and microorganisms decomposes the tissue surrounding the fibers before extensive damage is done to the fibers themselves. Judgment and experience are necessary to accomplish retting without damaging the quality of the fiber.

Uses

Fine flax may be used for apparel fabrics (called linens), such as damasks, sheetings, and thread; coarse types are used for twines, canvas, and bags. Also, high grade paper, including cigarette paper, is made from flax. Hemp has a wide range of uses from sacking to carpets to upholstery to cordage. Jute is used primarily for sacks and bags. It is also used as a backing for carpets and in twine and rope. Compared to the other fibers, jute is relatively weak and perishable. It rapidly deteriorates when exposed to moisture and dampness. Kenaf is used for coarse materials, namely sackings, ropes, and twine. Whole dry stems of kenaf have been used for pulp to make paper.

Man-made fibers have taken the market positions of many of the bast fibers. Rayon, polyester, and nylon are used for the apparel and household uses formerly occupied by flax and hemp. Polypropylene is now widely used for carpets, carpet backing, and bagging.

LEAF FIBERS

History

Abaca is a relative of the banana and is native to the Philippines. Pigafetta, a companion of Magellan, noted in 1520 that the natives of Cebu wore clothing made from abaca. Exports of abaca from the Philippines were begun early in the 19th century. While attempts have been made to introduce *M. textilis* culture into other moist tropical areas, the Philippines remain the major producer. Well over 90% of the world's abaca fiber is produced in the Philippines each year.

Agave spp. (which resemble the century plant) are native to the New World. The Mayans used fiber from *Agave*, and *A. sisalana* was introduced from Mexico through Florida into East Africa in the latter part of the 19th century. Production in Africa is concentrated in Tanzania, Angola, Kenya, and Mozambique. Henequen, *A. fourcroydes*, is grown almost entirely in Yucatan.

Structure

Leaf fibers are from the vascular systems of the leaves and leaf sheaths. The individual fiber cells elongate during development of the leaf and are pointed on both ends. The secondary wall develops as in the bast fibers, but there is a larger lumen. Individual abaca fibers are 2 to 12 mm in length, and those of *Agave* are 2 to 5 mm. Widths are 15 to 30 μm. The leaf fibers in general are stronger but more coarse and less flexible than bast fibers. The fibers of commerce are aggregates of individual fiber cells.

Fiber Quality

Abaca fiber is graded on the basis of cleaning and color. Length, color, cleanliness, and fiber alignment are factors in grading sisal and henequen.

Surrounding tissue must be removed from the fibers. Decortication is the process of crushing and scraping away the extraneous tissue from the fibers. Following decortication, fiber aggregates remain Abaca aggregates or strands are 1.5 to 3 m, and those of sisal and henequen are 1 to 2 m in length.

Uses

Abaca is used for rope and twine. Its resistance to salt water, strength, and low amount of swelling when wet make it particularly useful in marine ropes. Sisal and henequen are used primarily for binder and baler twines.

The man-made fibers are being used more and more in recent times for rope and twine. Thus, the use of leaf fibers is also declining.

SUMMARY

Fibers may be obtained from the outgrowth of single cells on the seed as in cotton; from phloem fibers of some herbaceous dicotyledonous plants as in flax, hemp, jute, and kenaf; and from vascular strands in leaves and leaf sheaths of monocotyledonous plants as in abaca, sisal, and henequen. Cellulose is the major constituent of plant fibers, ranging from more than 90% in cotton to 71% in flax. Plant fibers are used in a wide range of products from lightweight apparel fabrics to bagging, rope, and twine.

PHOTO CREDITS

Fig. 2.1 & 2.2—Courtesy of Cotton Quality Laboratory, Science and Education Administration, USDA; Fig. 2.3, 2.4, 2.7, & 2.8, from Rowland et al. (1976); Fig. 2.5, & 2.6 courtesy of International Institute for Cotton.

SUGGESTED READING

Anonymous. 1965. The classification of cotton. USDA Misc. Publ. 310.

————. 1976. Cotton counts its customers. National Cotton Council of America, Memphis, TN.

————. 1978. ASTM Method D-1445. Standard method of test for breaking strength and elongation of cotton fibers (flat bundle method). Annual Book of ASTM Standards, Part 33, Am. Soc. for Testing and Materials, Philadelphis, PA.

————. 1978. ASTM Method D-1447. Standard method of test for length and length uniformity of cotton fibers by Fibrograph measurement. Annual Book of ASTM Standards, Part 33, Am. Soc. for Testing and Materials, Philadelphia, PA.

————. 1978. Summary of cotton fiber and processing test results, crop of 1977. USDA Agric. Marketing Service, Cotton Division, Memphis, TN.

Beasley, C. A. 1975. Developmental morphology of cotton flowers and seed as seen with the scanning electron microscope. Am. J. Bot. 62:584–592.

Berger, Josef. 1969. The world's major fibre crops: Their cultivation and manuring. Centre d'Etude de l'Azote, Zurich.

Brown, H. B., and J. O. Ware. 1958. Cotton, 3 ed., McGraw-Hill Book Co., Inc., NY.

Dempsey, J. M. 1975. Fiber crops. Univ. Presses of Florida, Gainesville.

Doberczak, A., St. Dowgielewiecz, and W. Zurek. 1964. Cotton bast and wool fibers [Transl. from the 1958 Polish ed.]. Office of Tech. Serv., U.S. Dep. of Commerce. OTS 60-21551.

Evans, R. B. 1973. The world's cottons: A summary of cotton fiber and processing test results. USDA Foreign Agric. Serv. FAS M-250.

Gatewood, L. B., Jr. 1973. The energy crisis: Can cotton help meet it? National Cotton Council of America, Memphis, TN. [as updated by Randall Jones (1977), personal communication].

Hearle, J. W. S., and R. H. Peters. 1963. Fibre structure. Butterworth's, Manchester, England.

Kirby, R. H. 1963. Vegetable fibres. Interscience Publ., Inc., NY.

Linton, G. E. 1966. Natural and manmade textile fibers. Duell, Sloan and Pearce, NY.

Mukherjee, R. R., and T. Radhakrishnam. 1972. Long vegetable fibres. Text. Progr. 4(4):1–75.

Pearson, N. L. 1955. A study of cotton fiber perimeters as calculated from Arealometer values at low and high compression. Text. Res. J. 25:124–136.

Ramey, H. H., Jr. 1974. The qualities of cotton produced in the U.S. Modern Text. 55(10):58, 60, 68

————, R. Lawson, and S. Worley, Jr. 1977. Relationship of cotton fiber properties to yarn tenacity. Text. Res. J. 47:685–691.

Robinson, B. B., and F. L. Johnson. 1953. Abaca—A cordage fiber. USDA Agric. Monograph 21.

Rowland, S. P., M. L. Nelson, C. M. Welch, and J. J. Hebert. 1976. Cotton fiber morphology and textile performance properties. Text. Res. J. 46:194–214.

Schubert, A. M., C. R. Benedict, J. D. Berlin, and R. J. Kohel. 1973. Cotton
fiber development-kinetics of cell elongation and secondary wall thickening. Crop Sci. 13:704–709.

Stewart, J. M. 1975. Fiber initiation on the cotton ovule (*Gossypium
hirsutum*). Am. J. Bot. 62:723–730.

Waterkeyn, L., E. de Langhe, and A. A. H. Eid. 1975. In vitro culture of fertilized cotton ovules. II. Growth and differentiation of cotton fiber. La
Cellule 71:41–54.

Zylinski, T. 1964. Fiber science [Transl. from the 1958 Polish ed.]. Office of
Tech. Serv., U.S. Dep. of Commerce. OTS 60-21550.

Chapter 3

Forage Crops

J. E. MOORE

University of Florida
Gainesville, Florida

The efficiency of forage utilization by livestock is dependent upon a large number of factors, including quality and quantity of available forage, production potential of the animal, and the amount and nature of supplemental feeds. The livestock producer must understand each of these factors in order to make wise management decisions.

This chapter provides some insight into animal science and some perspectives about forage quality and utilization of forages by ruminant livestock. Although economic efficiency is of utmost practical importance, it will be considered only briefly, because cost-price ratios are highly variable.

FORAGE QUALITY

Forage quality is much more variable than grain quality. There are differences among plant genotypes, seasons of the year, and stages of maturity. Because of this variability, it is necessary to know the quality of forage being consumed in order to make objective feeding and management decisions.

Definition of Forage Quality

Definitions of forage quality are numerous and varied. The terms leafy, fine-stemmed, green, sweet-smelling, high-protein, low-fiber, and palatable have been used to describe high quality forages. In recent years, agronomists and animal scientists have learned to rely upon biological responses of ruminant animals for accurate measures of forage quality. It is, after all, the animal, rather than the human, which makes the ultimate evaluation of forage quality. The most useful definitions of forage quality are in terms of animal performance, or voluntary intake of digestible energy (DE).

Table 3.1—Hypothetical comparison of two forages fed to 300 kg (660-lb) yearling beef steers for 90 days.

	Forage	
Item	A	B
Average daily gain, kg (lb)	0.6 (1.32)	1.2 (2.64)
Average daily forage, kg (lb):		
Dry matter offered	11.0 (24.23)	13.8 (30.39)
Dry matter refused	1.4 (3.08)	1.8 (3.96)
Dry matter consumed	9.6 (21.15)	12.0 (26.43)
Feed efficiency:		
Dry matter consumed per unit of gain	16.0	10.0
Total forage required for 90 days, MT (tons):		
Per steer	0.99 (1.09)	1.24 (1.37)
Per 6 steers	5.94 (6.54)	7.45 (8.20)

Animal Performance

Forage quality is best defined in terms of animal performance: for example, average daily gain by growing steers, or average daily milk production by dairy cows. Some qualifications are necessary, however: 1) the forages being compared must be the only sources of energy and protein, 2) the amount offered must exceed consumption by 5 to 15% (i.e., free-choice feeding), and 3) the animals must have potentials for production (gain or milk).

Table 3.1 presents a hypothetical comparison of two forages offered free-choice to beef steers. The steers offered forage B gained at a faster rate. Thus, forage B had higher quality than forage A. Forage B was consumed at a higher rate, and was more efficiently used, than was forage A.

There are practical difficulties in using animal performance to measure forage quality. An accurate measure of gain by steers can be obtained in no less than 90 days and with a minimum of six animals per forage. Thus, a large quantity of forage is required (Table 3.1). If animals of a different class, age, weight, or condition had been used, the absolute results (Table 3.1) would undoubtedly have been different, although the relative differences should have been about the same. If mature beef cows were used, however, their low potential for weight gain might have resulted in no apparent differences between the forages.

Voluntary Intake of Digestible Energy

The voluntary intake of DE is a definition of forage quality that is frequently useful. Calculation of DE intake is based upon measurements of voluntary intake and digestibility of forage energy. Voluntary intake measurements are made by offering the forage alone and free-choice. Fecal excretion is determined by either collecting the total

Table 3.2—Example calculation of voluntary intake of digestible energy.

Dry matter consumed/day†	1.2 kg (2.64 lb)
Forage gross energy concentration	4.4 Mcal/kg (2.00 Mcal/lb)
Gross energy consumed/day	5.28 Mcal
Dry matter excreted in feces/day‡	0.5 kg (1.10 lb)
Feces gross energy concentration	4.7 Mcal/kg (2.13 Mcal/lb)
Gross energy excreted/day	2.35 Mcal

Energy digestibility:

$$\frac{\text{intake-excretion}}{\text{intake}} \times 100 = \frac{5.28 - 2.35}{5.28} \times 100 = \frac{2.93}{5.28} \times 100 = 55.5$$

Voluntary intake of digestible energy:

$$\frac{\text{intake} \times \text{digestibility}}{100} = \frac{5.28 \times 55.5}{100} = 2.93 \text{ Mcal/day}$$

† Voluntary intake; mean of 7 days following a 14-day preliminary period; sheep data.
‡ Mean of 7 days collection during intake-measurement period.

excretion or using some means of estimating it. Both forage and feces are analyzed for gross energy concentration (energy of combustion), and calculations are made of gross energy consumption and excretion, energy digestibility, and intake of DE (Table 3.2). In effect, DE intake is the product of intake times digestibility.

Table 3.2 shows that the concentration of gross energy in feces is greater than that in forage. Fecal energy is increased because, in addition to undigested feed, feces contains metabolic protein or fat produced by the liver, pancreas and intestinal mucosa, and by intestinal microflora. The higher energy concentration in feces does not, however, reflect the extent of digestion. Fecal excretion of gross energy (2.35 Mcal) was less than the daily intake of gross energy (5.28 Mcal) by 2.93 Mcal. The latter value is the amount of energy which was "apparently digested," that is, which disappeared during the passage of feed through the entire digestive tract. Digestible energy is the term applied to this quantity of energy, but it represents the energy which is absorbed plus that lost as heat and methane from microbial fermentation. Further, the metabolic excretions of protein and fat make "apparent" DE values less than the "true" values. Even though DE is subject to several sources of variation in the intestinal tract, it is estimated with excellent precision and it is highly repeatable.

Studies of DE intake are similar to those of animal performance, but the length of the feeding period may be shortened from 90 to 21 days. Sheep, instead of steers, may be used to decrease the amount of feed required. Less than 200 kg (440 lb) of forage would be required for an intake trial with sheep, as compared to about 7,000 kg (15,400 lb) for a gain trial with steers. Because sheep are ruminants, results obtained with them may be applied, with caution, to other ruminants.

Voluntary intake of DE is an acceptable measure of forage quality because it is closely related to animal performance when the animal in question has the potential to gain or produce milk. A certain minimum DE intake is required just to maintain the weight and condition of the animal (Maintenance Requirement, Fig. 3.1). As DE intake increases above maintenance, the daily rate of production increases up to the point of animal potential. Since DE intake and rate of animal perform-

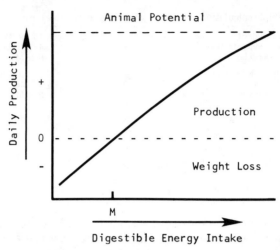

Fig. 3.1—Relationship between daily digestible energy intake and production per animal (M = maintenance requirement).

ance are closely related, energy requirements may be expressed in terms of DE.

Figure 3.2 illustrates the effects of forage maturity upon energy digestibility and voluntary intake of DE. As maturity increased from 4 to 8 weeks, both digestibility and DE intake tended to decrease linearly. Although there was relatively little decrease in energy digestibility from 8 to 10 weeks, there was a continuing decrease in DE intake. Thus, some factors, such as protein, may affect voluntary intake without affecting digestibility to the same extent.

Application of Forage Quality Data

Published DE intake values may show relative differences among forages for a single experiment, or the average of several experiments. Such data can be useful in illustrating principles of forage management and differences among species. However, many "non-forage" factors influence observed DE intake values, such as age, sex, weight, condition, health, and genetic potential of the animal; environmental temperature and humidity; and management practices. These "nonforage" factors affect the production response of animals under practical conditions, either through effects upon intake or metabolism. Therefore, published data on forage quality should not be applied by livestock managers without making certain that the values are appropriate for the specific forage-animal-environment combination.

Determinants of Forage Quality

The quality of a forage is determined by its chemical and physical characteristics. The interaction of these characteristics with the mechanisms of digestion, metabolism, and control of voluntary intake

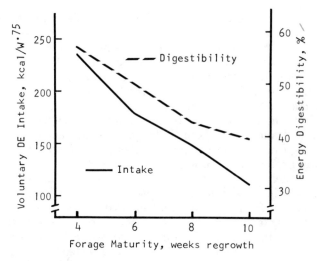

Fig. 3.2—Effect of forage maturity upon voluntary digestible energy (DE) intake ($W^{0.75}$ = kg metabolic body weight) and energy digestibility by sheep ['Suwannee' bermudagrass, *Cynodon dactylon* (L.) Pers., hay]. Adapted with permission from Golding et al. (1976).

determine the level of DE intake, and animal performance, which can be achieved. An understanding of forage quality requires knowledge of 1) forage composition, and 2) ruminant nutrition.

Forage Composition

The constituents of forages can be divided into two main categories: 1) those existing in the cell contents (e.g., protein, sugar, and starch), and 2) those which make up the structural components of the cell wall (e.g., cellulose, hemicellulose, and lignin). Structural carbohydrates are of particular importance because their digestion is dependent upon enzymes produced by gastrointestinal microorganisms. On the other hand, non-structural carbohydrates are readily digested by enzymes of both animal and microbial origin. The physical organization of forage cells into various tissues is quite complex (Fig. 3.3). The chemical constituents of forages are not distributed uniformly among different plant organs and tissues, and wide differences exist among forages in both composition and physical structure.

Several schemes of analysis have been developed to describe the gross composition of forages and other feeds. Such analyses are necessary in order to formulate rations for specific animals. In addition to gross constituents, feeds and forages contain many constituents which, although in low concentration, have important effects upon forage quality. Some of these constituents are vitamins and required minerals, but others have anti-quality effects.

A. *Gross Composition*. Figure 3.4 shows the major chemical constituents of forages and how they are fractionated by two systems of analysis: 1) the proximate analysis and 2) the Van Soest analysis.

1. Proximate analysis—Since the mid 1800's, proximate analysis has been widely used in the evaluation of feedstuffs. The six components of a complete proximate analysis are shown in Fig. 3.4. Moisture is the loss in weight when samples are dried at 105 C and ash is the residue remaining after burning at 600 C. Crude protein is determined by analyzing for N by the Kjeldahl method, and multiplying by 6.25. Ether extract is the total of all compounds which can be extracted with hot diethyl-ether. Crude fiber is the organic matter which is insoluble in weak acid and weak alkali. Nitrogen-free extract is calculated by subtracting from 100 the sum of the other five components.

Crude protein and crude fiber have been widely used to classify feeds, and their use has made it possible to formulate diets. Forages have been defined, with some exceptions, as those feeds having more than 18% crude fiber in the dry matter, as opposed to concentrates, defined as having less than 18% crude fiber. Table 3.3 presents the crude protein, crude fiber, and DE concentrations in a variety of feeds. The concentration of DE tends to be higher in those feeds having low crude fiber and high protein because, generally, fiber is not highly digestible. In general, there is a negative relationship between crude fiber and crude protein in feeds with more than 18% crude fiber (e.g., forages and byproducts), but considerable variation exists (Fig. 3.5).

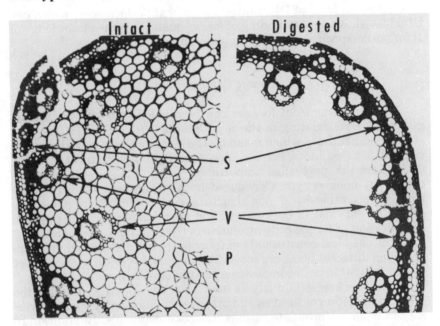

Fig. 3.3—Cross sections of stem of Pangola digitgrass (*Digitaria decumbens* Stent.; 4-week regrowth) illustrating types and distribution of tissues (S = sclerenchyma, V = vascular bundles, P = parenchyma). "Intact" section is from untreated stem, and "digested" section is from residue remaining after digestion by rumen microorganisms in vitro. Lignified vascular bundles and sclerenchyma resisted digestion but parenchyma was almost completely digested. Cell walls, which appear in the photomicrograph, are constituents of neutral-detergent fiber, whereas cell contents are soluble in neutral-detergent reagent.

Table 3.3—Crude protein, crude fiber, and digestible energy concentration of some feeds. †‡

Feed class and name	CP	CF	DE	Feed class and name	CP	CF	DE
Dry forages and roughages				Energy feeds			
Alfalfa:				Barley grain	13	6	3.8
—dehydrated	19	26	2.5	Beet pulp, dried	10	21	3.2
—hay, vegetative	25	20	2.6	Citrus pulp, dried	7	14	3.4
—hay, midbloom	18	31	2.4	Corn grits (hominy)	12	6	3.9
—hay, mature	14	38	2.2	Corn grain, grade 2	10	2	4.1
Barley hay	9	26	2.5	Oats grain, grade 2	12	12	3.3
Barley straw	4	42	1.8	Rice bran	15	12	3.5
Bermudagrass hay	9	30	2.2	Rye grain	13	2	3.8
Blue grass hay	12	29	2.4	Sorghum, Milo, grain	12	2	4.1
Brome hay	12	32	2.3	Sugarcane molasses	4	0	3.2
Canary grass,				Wheat bran	18	11	2.9
Reed, hay	11	34	2.3	Wheat grain	14	3	3.9
Clover hay:							
—Ladino	23	19	2.5				
—Red	15	30	2.6				
Corn cobs	3	36	2.2	Protein supplements			
Corn fodder	9	26	2.8	Coconut meal	22	13	3.7
Corn stover	6	37	2.7	Corn distillers:			
Cottonseed hulls	4	48	2.5	—grains	30	13	3.6
Fescue hay	10	31	2.5	—solubles	29	4	4.0
Oats hay	9	31	2.4	Corn gluten meal	47	4	3.8
Oats straw	4	41	1.9	Cotton seed meal	45	13	3.0
Orchardgrass hay	10	34	2.4	Flax (linseed) meal	3	10	3.5
Soybean hulls	14	39	1.9	Peanut meal	49	12	3.9
Timothy hay:				Safflower meal	22	34	2.1
—vegetative	12	33	2.6	Sesame meal	52	5	3.6
—late bloom	8	32	2.3	Soybean meal	52	7	3.5
Wheat straw	4	42	1.7	Sunflower meal	50	12	2.9

† Reprinted from National Research Council (1975), with the permission of the National Academy of Sciences, Washington, D.C.
‡ CP = Crude protein, % of dry matter; CF = crude fiber, % of dry matter; DE = digestible energy, Mcal/kg of dry matter.

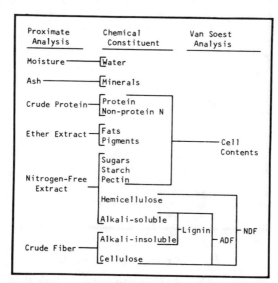

Fig. 3.4—Major chemical constituents of forage and their fractionation by two systems of laboratory analysis (ADF = acid-detergent fiber; NDF = neutral-detergent fiber, an estimate of cell walls). Adapted with permission from Van Soest (1967).

68   J. E. MOORE

Crude fiber is generally thought to be less digestible than N-free extract, but with forages there is no clear distinction (Table 3.4). In some cases, N-free extract may actually be less digestible than crude fiber because most of the hemicellulose and part of the lignin are included in N-free extract (Fig. 3.4). Hemicellulose is a structural carbohydrate and its digestibility is similar to that of cellulose. Lignin is associated with cellulose and hemicellulose in the cell wall. Lignin is usually indigestible, but its major effect is to inhibit the digestibility of cellulose and hemicellulose. Therefore, when hemicellulose and lignin percentages are high, the digestibility of N-free extract is depressed to values similar to or lower than those for crude fiber (Table 3.4). These data show that the proximate analysis does not characterize the quality-related properties of forage carbohydrates.

2. *Van Soest analysis*—Peter J. Van Soest developed an alternative to the proximate analysis while working at the USDA laboratories in Beltsville, Maryland, during the 1960's. His methods recognized the distinction between cell walls and cell contents (Fig. 3.4). The most important procedure involves treating a forage sample with a neutral-detergent solution: the solubles are primarily the cell contents, and the insoluble residue (neutral-detergent fiber) is an excellent estimation of the total structural, or cell-wall, constituents (cellulose, hemicellulose, and lignin). Neutral-detergent fiber varies from 10% in corn grain to

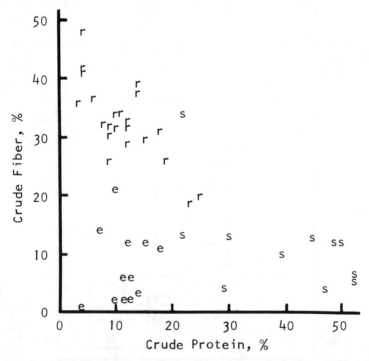

Fig. 3.5—Crude protein and crude fiber content of some feeds (from Table 3.3); r = dry forages and roughages, e = energy feeds, s = protein supplements (values are expressed as percent of dry matter).

Table 3.4—Average digestibility of crude fiber (CF) and N-free extract (NFE) of various feeds.†

Feeds	No. of feeds	Digestibility		Feeds with NFED ≤ CFD‡	
		CF	NFE	No.	%
		— % —			
Dry roughages	110	52	59	43	39
Green roughages	61	64	76	12	20
Silages	25	58	65	7	28
Concentrates	88	53	78	9	10
All feeds	284	56	69	71	25

† Adapted with permission from Crampton and Maynard (1938).
‡ Feeds having NFE digestibility (NFED) less than or equal to CF digestibility (CFD).

80% in straws and tropical grasses. Acid-detergent fiber is an insoluble residue, like neutral-detergent fiber, but does not include all cell-wall constituents because hemicellulose is soluble in the acid-detergent solution. Acid-detergent fiber varies from 3% in corn grain to 40% in mature forages and 50% in straws. Acid-detergent fiber values are slightly higher than are those for crude fiber because all the lignin and some ash is included in the former. Van Soest developed an improved lignin analysis by treating acid-detergent fiber with 72% sulfuric acid or permanganate.

Neutral-detergent fiber is most important because it estimates that fraction of forage which, if it is to be metabolized by the animal, must first be degraded by gastrointestinal microorganisms. It is a better measure of forage "fiber" than crude fiber, but it is not uniformly digestible. In fact, the digestibility of energy is often closely related to the digestibility of neutral-detergent fiber in many forages, especially grasses. Microscopic studies (Fig. 3.3) show that the walls of some cells undergo rapid and complete degradation by microbial and mechanical actions in the gastrointestinal tract, while walls of other cells are more resistant to degradation. A more thorough understanding of these phenomena will lead to improvement in forage quality and forage utilization.

B. *Vitamins and Minerals.* Forages are important sources of Vitamins A and E, and the essential minerals Na, K, Ca, P, Mg, S, Cu, Co, Zn, Fe, Mn, Mo, I, and Se. There are no "group" methods of analysis which permit rapid evaluation of the vitamin or mineral value of feeds. Determination of the adequacy, deficiency or toxicity of vitamins or minerals in feeds is often based upon observations of animals for specific signs or symptoms. However, subclinical amounts (deficient or toxic) may depress intake, digestibility, or animal performance even though signs of acute problems are not apparent. Certain animal tissues may be analyzed for nutrient concentration in order to determine the status of an animal with respect to a given nutrient.

C. *Anti-quality Factors.* The levels of various anti-quality factors must not exceed certain limits. Molds, dust, weeds, alkaloids, tannins, nitrates, and cyanides may cause toxic reactions, cause the feed to be unpalatable, or decrease microbial activity in the rumen. The net result may be decreased feed intake, digestibility, and animal performance. Analyses for these factors in feeds are specific, and may be time-consuming or expensive. Tissue analyses are sometimes used to diagnose these problems.

Ruminant Nutrition

Nutrition may be defined as the acts or processes whereby the food ingested by an animal is used for purposes of maintenance of life, growth, reproduction, lactation, and work. In order to understand the relationship between forage composition and forage quality, it is necessary to review important aspects of ruminant nutrition such as digestion, metabolism, and voluntary-intake control.

A. *Digestion.* Digestion involves preparation of food for absorption. Diets of farm animals consist largely of macromolecules such as proteins, pectins, starches, cellulose, and hemicellulose. These complex compounds must be digested to simple compounds before they can be absorbed into the bloodstream. Many digestive enzymes are produced by the stomach, pancreas, and small intestine, but none can digest cell wall carbohydrates like cellulose and hemicellulose.

Cellulose and hemicellulose are digested in animals only by the enzymes of microorganisms in the gastrointestinal tract. Herbivores have enlarged compartments in the gastrointestinal tract which favor the development of a microbial population. The herbivorous non-ruminants, including horses, rabbits, and elephants, have an enlarged colon and caecum, and major microbial activity occurs after the ingested feed has been subjected to digestion by animal enzymes in the stomach and small intestine. Ruminants, such as cattle, sheep, goats, deer, camels, llamas, and buffaloes, have enlarged and compartmented stomachs. Microbial digestion in ruminants occurs in certain compartments of the stomach, primarily, but some microbial action does occur in the colon and caecum.

Stomach function, microbial digestion, and fermentation are described in the following paragraphs.

1. *Stomach function*—A characteristic of ruminants is that they regurgitate a portion (bolus) of previously consumed feed, chew the bolus for a few moments, and reswallow the bolus along with saliva. This process is called rumination. More than 30,000 jaw movements have been recorded for a cow in a 24-hour period. Mechanical fragmentation of forage occurs during rumination because of chewing and agitation. The mechanical degradation creates sites for attack by rumen microorganisms. Microbial digestion, in turn, weakens the forage structure so that the tissues are more easily fragmented.

There are four compartments in the stomach of domestic ruminants, the rumen, reticulum, omasum, and abomasum (Fig. 3.6). The rumen is the largest and may have a capacity of about 150 liters (40 gal) in a mature dairy cow. It is subdivided by strong muscles that contract periodically, mixing and agitating its highly fluid contents.

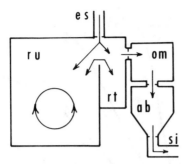

Fig. 3.6—Representation of the flow of digesta through the stomach compartments of domestic ruminants (es = esophagus, ru = rumen, rt = reticulum, om = omasum, ab = abomasum, si = small intestine).

When forage is consumed as pasture and long hay, a thick mat of forage floats on top of the rumen fluid. The processes of mechanical and microbial degradation gradually reduce the size of forage particles until they settle to the bottom of the rumen and are transferred to the reticulum.

The reticulum has been called the "hardware stomach," because wires, nails, and other solid materials sometimes collect in it. The reticulum has important functions in controlling movement of feed particles, such as forming the bolus which is ruminated and passing small, dense residues to the omasum. The opening between reticulum and omasum controls the size of forage particles passing through it, and retains particles in the rumen and reticulum until they are small enough to pass to the omasum. Retained particles pass back and forth freely between rumen and reticulum, and may be ruminated. Retention influences voluntary intake.

The omasum has been called "many-plies" in reference to the internal folds that appear as pages of a book when it is opened. The internal surface area is high compared to volume, and the omasum dehydrates the material passing through it from reticulum to abomasum.

The abomasum is often called the "true stomach" because it secretes hydrochloric acid and pepsin, a protein-digesting enzyme. Of the four stomach compartments in domestic ruminants, the abomasum is the only one that synthesizes a digestive enzyme, but pepsin has no action on structural carbohydrates.

2. *Microbial digestion and fermentation*—The microorganisms in the rumen and reticulum include both bacteria and protozoa. A large number of different species have been identified and most act on one specific substrate, such as cellulose, starch, or protein, among many others. The composition of the diet determines the type of microbial population and the rate of microbial digestion (Fig. 3.7). If the diet is high in starch, for example, starch-digesting organisms grow rapidly and starch is quickly digested. If the diet is a forage with a high percentage of structural carbohydrates, the microbial popula-

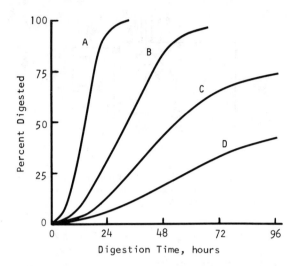

Fig. 3.7—Hypothetical digestion curves of several substrates by rumen microorganisms (A = purified starch, B = delignified, purified cellulose, C = cell walls of high-quality forage, D = cell walls of low-quality forage).

tion is different, and digestion occurs more slowly. If forage lignin increases, digestion of structural carbohydrates decreases.

Digestion is only a part of the microbial activity in the rumen and reticulum (Fig. 3.8). Products of digestion, such as amino acids or monosaccharides, are seldom found in rumen fluid because microorganisms ferment these products and produce adenosine triphosphate (ATP) for their own growth. Fermentation is anaerobic and incomplete, however, and energy-rich end-products pass from the microbial cells into the rumen fluid. The major rumen fermentation end-products are the volatile fatty acids (VFA): acetic, propionic, and butyric acids. The ratio of acetic to propionic acid is influenced by the type of carbohydrate fermented. Structural or cell wall carbohydrates produce a high ratio of acetic to propionic acid, while starch produces a low ratio. Fermentation also produces waste heat and methane which cannot be used by the animal for production. An important function of saliva is to provide sodium bicarbonate to buffer the VFA. Rumen fluid pH is normally 6.5 to 6.8 when the diet is primarily forage.

Protein and natural non-protein N (NPN) compounds are also fermented by rumen microorganisms, producing VFA and ammonia (Fig. 3.8). Ammonia is the main source of N for rumen microbial growth. Some microorganisms require ammonia because they cannot use protein or amino acids. Synthetic NPN compounds, such as urea, biuret, and ammonium salts may be fed because the rumen microorganisms convert the N to ammonia and then to microbial protein. Readily digested soluble carbohydrates and sulfur are required for optimum use of NPN because they support rapid microbial growth and protein synthesis. Structural carbohydrates are not digested rapidly enough for maximum NPN utilization.

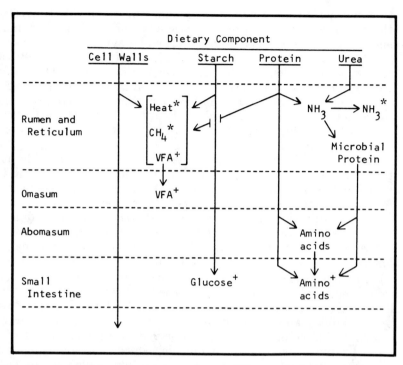

Fig. 3.8—Digestion and fermentation in ruminants (VFA = volatile fatty acids; + = products which are absorbed and utilized in animal metabolism; * = products which are absorbed but excreted).

Microbial digestion and fermentation, and microbial protein synthesis, are dependent upon the availability of nutrients required by the microorganisms, especially N. The rates of microbial carbohydrate digestion are decreased when the percentage of crude protein is below about 7% of the diet dry matter. Rumen microorganisms have definite nutritional requirements for many minerals as well as for energy and N. Deficiencies of Ca, P, Mg, S, and Co have been described. Such deficiencies result in reduced digestion of structural carbohydrates and/or synthesis of protein.

B. Metabolism. The process of metabolism produces energy and protein in the animal.

1. Energy—The VFA produced by microbial fermentation provides up to 80% of the energy used by adult ruminants consuming forage. Growing and adult ruminants have definite requirements for VFA, and normal growth and development are inhibited when rumen function is impaired by artificial means. Acetic acid is used directly for ATP production and fat synthesis by several tissues, including the mammary gland. In lactating dairy cows, low levels of acetic acid are associated with low percentages of fat in milk. Propionic acid is converted by the liver to blood glucose, and this provides most of the glucose in the blood of forage-fed ruminants. Glucose may be used for ATP production by many tissues, and as the main precursor for lactose synthesis by the mammary gland. Butyric acid is converted by the rumen wall to β-hydroxy butyric acid, which is used by muscles and other tissues for ATP production and fat synthesis.

The ratio of acetic acid to propionic acid influences the efficiency of energy metabolism. A decreased proportion of acetic acid, such as when grain is fed, increases the efficiency of energy utilization because of decreased heat loss during metabolism. Conversion of grain energy to animal product is more efficient than that of forage energy because of both higher digestibility of grain energy and lower loss of energy during metabolism.

2. *Protein*—Microbial protein synthesized during rumen fermentation is well digested by the animal in the abomasum and small intestine (Fig. 3.8). Amino acids are absorbed and used in protein synthesis. Microbial amino acids generally include the "essential" amino acids which the animal cannot otherwise synthesize in adequate quantities for growth and production. For this reason, nutritionists are generally concerned with the percentage of crude protein in the ruminant diet and not with the percentages of individual amino acids.

Ruminants can adapt to forages with a wide range of N concentrations. Some rumen ammonia is absorbed and converted to urea by the liver, and urea is transferred to the blood. Blood urea may be removed by the kidney and excreted in the urine, or recycled back into the rumen via saliva or direct transport across the rumen wall. Recycled urea is rapidly converted to ammonia, which can be used by rumen microbes. Ruminants can adapt to low-protein forages by, in part, increasing the efficiency of N recycling and reducing urinary urea excretion. Low-protein forages may be the only feed available during the dry season or winter on many grasslands of the world. When high-protein forages are consumed, the liver and kidney prevent ammonia toxicity by increasing urinary urea excretion. Because excretion is a waste of dietary N, there can be little justification for providing ruminants with a greater N intake than they require.

C. *Control of Voluntary Intake.* An understanding of the factors controlling forage intake is essential to an understanding of forage quality. Control of voluntary intake in ruminants is a complex of animal-related factors, diet-related factors, and their interactions. The two major theoretical control mechanisms are: (1) chemostatic, and (2) distention (Fig. 3.9).

1. *Chemostatic mechanism*—For a given animal, maximum voluntary DE intake (appetite) may be determined by some chemostatic mechanism. This mechanism may be sensitive to the blood concentration of certain metabolites, such as VFA, glucose, amino acids, or fats. When this mechanism is the one controlling intake, an increase in DE concentration of the diet will decrease dry matter intake, but DE intake will not change (Fig. 3.9). The DE concentration of most forages is below the range where the chemostatic mechanism operates. Therefore, few forages are consumed at rates which provide the maximum DE intake.

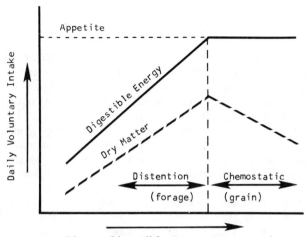

Fig. 3.9—Effect of increasing digestible energy concentration of the diet upon the daily voluntary intake of digestible energy and dry matter, and illustrating the operation of two theoretical intake control mechanisms. Adapted with permission from Montgomery and Baumgardt (1965).

2. *Distention mechanism*—Distention mechanism most likely operates with forages. It involves sensitivity to the degree of distention or "fill" of the gastrointestinal tract. There are stretch receptors in the rumen that detect the degree of distention and send signals to intake control centers. The animal will eat until the rumen and reticulum are distended, and will not resume eating until they are partially emptied of forage residues. The rumen and reticulum empty through two simultaneous processes: 1) digestion and absorption, and 2) passage of undigested particles to the omasum. Thus, intake is inversely related to the length of time forage residues are retained in the rumen and reticulum: the longer the retention time, the lower the intake.

Digestibility alone does not control forage intake; two forages having the same digestibility may not have the same retention time. Mechanical degradability of forage cell-wall particles may be important in controlling intake, because the faster particles are degraded, the shorter their retention time. Grinding and pelleting of forages decrease retention time and increase intake. Leaves may also be degraded faster than stems, but the amount of leaves is not always closely related to intake. Animals fed low-protein forages consume less than their DE requirement, are generally thin, and do not appear to be "full." Protein deficiency may reduce intake either by affecting the mechanisms of intake control or by decreasing rates of microbial digestion in the rumen.

FORAGE UTILIZATION

The daily rate of animal production is one of the most important considerations in any livestock enterprise. With increasing daily production, whether as meat, milk, or wool, the cost per unit of product generally decreases. This may occur because 1) the cost of overhead, labor, taxes, etc., may be constant on a daily per animal basis, and 2) the biological efficiency (product/feed) is generally greater with increasing rates of production. Livestock research has attempted to increase daily production of domesticated ruminants by improving the genetic potential of animals and by improving feeding and management practices.

The ultimate measure of biological efficiency in livestock production may be yearly animal production per unit of land area. This measure is determined by daily production per animal and the area of land required to produce the feed consumed per animal. With most livestock enterprises, however, a major objective of management will be to maximize production per animal.

Forages provide nearly three-quarters of the DE consumed annually by ruminants in the USA, and a higher proportion worldwide. In the USA, beef and milk production became increasingly dependent upon grain feeding in the two decades following World War II. Increased grain use resulted in higher rates of daily production with less feed, due to 1) higher intake of grain, 2) higher digestibility of grain energy, and 3) higher efficiency of grain energy utilization. As a result, the daily intake of DE and the efficiency of its use were higher than when forage was fed.

In much of the world, it is most economical to make maximum use of forages in ruminant production. This may be the case in some areas of the USA in the near future. Forages have been used extensively for the maintenance of mature breeding animals, such as non-lactating pregnant females, and this practice will continue. For the producing animal, such as the lactating dairy cow or the growing beef steer, efficient production on high dietary levels of forage will depend on application of knowledge of animal requirements and principles of forage utilization.

Feeding Systems

In this context, a feeding system has two components: 1) animal requirements, and 2) nutritional composition of the feeds used to meet these requirements. Both are expressed in comparable terms. The nutrient requirements of domestic animals and the composition of feeds have been summarized by the National Research Council (NRC) of the U.S. National Academy of Sciences. Appropriate NRC publications are listed at the end of this chapter. The requirement for a specific nutrient may be stated in two ways: 1) as a daily minimum quantity, or 2) as a minimum concentration in the total diet.

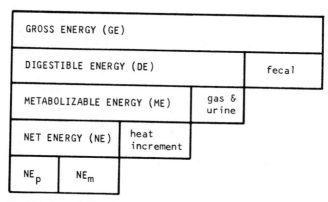

Fig. 3.10—Partitioning of dietary gross energy (NEp = net energy for production; NEm = net energy for maintenance). Adapted with permission from Crampton and Harris (1969).

Comparison of Feeding Systems

Since the mid-1800's a number of feeding systems have been proposed in Europe and the USA. Several systems in use in the USA will be discussed briefly.

A. *Protein.* An animal's requirements for protein may be given as grams of crude protein per day, and the crude protein concentration of feeds may be expressed as a percentage of the dry matter. In practice, digestible protein may be calculated from crude protein using formulas given by NRC. However, heat damage of feeds reduces digestible protein concentration. Heat damage may occur during artificial dehydration of forages and byproduct feeds such as citrus pulp, during ensiling of hay-crop forages, and during storage of baled hay which has not been dried properly.

B. *Energy.* Dietary energy may be partitioned into various fractions, including DE, metabolizable energy (ME) and net energy (NE) (Fig. 3.10). Requirements may be expressed as Mcal per day, and feed concentration as Mcal/kg of dry matter. Several systems in the USA are based upon NE, and permit measurement of the energy actually used in production. An English system based upon ME makes adjustments for differences in efficiency of ME utilization.

The major variation between forages with respect to energy is accounted for by differences in DE. The amount of NE for production which is available to a forage-fed ruminant is closely associated with DE intake. In turn, DE intake is related to intake and digestibility of forage energy.

For many years, the total digestible nutrient (TDN) system has been used in the USA as an energy feeding system. The TDN content of a feed is the sum of all the digestible organic components, weighted for their caloric value in metabolism. In practice, TDN is closely related to DE, and one may be calculated from the other according to the following formula:

Table 3.5—Major-nutrient requirements of selected ruminants.†

Animal	Body weight		Daily production‡		Daily requirements§						
					CP		DE	Ca		P	
	kg	(lb)	kg	(lb)	kg	(lb)	Mcal	g	(oz)	g	(oz)
Beef cattle											
Growing—finishing steers	200	(441)	0.90	(1.98)	0.61	(1.34)	16.3	23.0	(0.81)	18.0	(0.63)
	300	(661)	0.90	(1.98)	0.81	(1.79)	23.8	22.0	(0.78)	19.0	(0.67)
	300	(661)	1.30	(2.87)	0.83	(1.83)	26.4	29.0	(1.02)	23.0	(0.81)
	400	(882)	1.30	(2.87)	0.90	(1.98)	32.1	25.0	(0.88)	22.0	(0.78)
Mature cows											
dry, mid-pregnancy	500	(1,102)	--		0.42	(0.93)	17.2	13.0	(0.46)	13.0	(0.46)
nursing calves	450	(992)	5.0	(11.0)	0.86	(1.89)	22.0	26.0	(0.92)	26.0	(0.92)
Dairy cattle											
Growing heifers	150	(331)	0.75	(1.65)	0.44	(0.97)	11.9	15.0	(0.53)	12.0	(0.42)
	300	(661)	0.75	(1.65)	0.64	(1.41)	19.8	24.0	(0.85)	18.0	(0.63)
	450	(992)	0.75	(1.65)	0.89	(1.96)	23.4	27.0	(0.95)	21.0	(0.74)
Mature lactating cows											
5.0% butterfat	450	(992)	10.0	(22.0)	1.45	(3.20)	31.8	47.0	(1.66)	36.0	(1.26)
4.5% butterfat	500	(1,102)	15.0	(33.0)	1.87	(4.12)	40.0	62.0	(2.19)	47.0	(1.66)
4.0% butterfat	600	(1,323)	20.0	(44.0)	2.29	(5.05)	48.1	76.0	(2.68)	57.0	(2.01)
3.5% butterfat	600	(1,323)	30.0	(66.0)	2.95	(6.50)	59.1	100.0	(3.53)	74.0	(2.61)
Sheep											
Ewes											
non-lactating	60	(132)	0.03	(0.07)	0.12	(0.26)	3.2	3.1	(0.11)	2.9	(0.10)
early lactation	60	(132)	-0.03	(-0.07)	0.24	(0.53)	6.6	11.5	(0.41)	8.2	(0.29)
replacements	40	(88)	0.12	(0.26)	0.13	(0.29)	3.6	6.1	(0.22)	3.4	(0.12)
Finishing lambs	30	(66)	0.20	(0.44)	0.14	(0.31)	3.7	4.8	(0.17)	3.0	(0.11)
	45	(99)	0.25	(0.55)	0.19	(0.42)	5.2	5.0	(0.18)	3.1	(0.11)

† Reprinted from National Research Council (1971, 1975, 1976), with the permission of the National Academy of Sciences, Washington, DC.
‡ Production values are in terms of weight changes, except for milk production by lactating beef and dairy cows.
§ CP = Crude protein; DE = Digestible energy (estimated for beef cattle as 4.4 Mcal/kg TDN).

$$\text{Mcal DE} = 4.4 \times \text{kg TDN}$$

C. *Vitamins and Minerals*. Vitamin and mineral requirements are stated simply in terms of minimum daily quantities, or minimum concentrations in the diet. Each vitamin and mineral in the diet must be expressed separately. Depending upon the specific nutrient, requirements may be in terms of grams, milligrams, or units required per day, or as percentage, parts per million, or units/kg required in the diet.

Nutrient Requirements of Various Ruminants

Nutrient requirements of animals are proportional to their size, physiological function, and daily rate of production. The NRC publication series, Nutrient Requirements of Domestic Animals, presents the most complete information available. These books are revised frequently, when new information is generated from research and practical experience. The requirements of a few selected ruminants for certain nutrients are presented to illustrate general principles (Table 3.5). Maximum daily production per animal should result when the intakes of all nutrients are in balance and at least equal to requirements. With forage as the only source of nutrients, nutrient intakes are often less than required for maximum production per animal.

In addition to nutrient requirements, certain other diet characteristics must be considered. A small amount of roughage is needed to avoid rumen dysfunction, digestive disorders, and decreased appetite. A sign of normal rumen function is the visual observation that the animal is ruminating, but steers on high-concentrate diets may be healthy even if they are not ruminating. However, rumen parakeratosis and liver condemnations are potential problems on diets which are very low in fiber. Dairy cow diets must have a minimum amount of fiber in order to maintain rumen acetic acid concentration and milk-fat percentage at normal levels.

Nutritional Characterization of Forages

The ultimate method of evaluating the quality of forage involves feeding it to an appropriate animal. If one wants to use forage quality information in formulating rations to be fed, however, it is necessary to predict in advance what the DE intake will be. This method involves obtaining a small sample of the forage to be fed and analyzing it in a laboratory.

Forages may be analyzed for nutrients and anti-quality factors by the same procedures used for grains and protein supplements. Such methods, however, are not completely adequate for the prediction of forage quality. A variety of special laboratory methods for evaluating

Table 3.6—Laboratory methods which have been used in predicting forage quality.†

Laboratory method	Quality measure‡
Chemical methods	
Availability index	DMD
Acid detergent fiber	DMD, DMI
Acid detergent lignin	DMD, DMI
Acid insoluble lignin	DMD
Cellulose	DMD
Crude fiber	DMD
Crude protein	DMD, CPD
Cell wall constituents (CWC)	DMD, DMI
CWC + anhydrouronic acid (AUA)	OMD
CWC + AUA + lignin	OMD
Lignin	DMD
Modified crude fiber	TDN
Methoxyl	DMD
Normal acid fiber	DMD
Summative equation	DMD
Sum of average in vivo digestible fractions	OMD
Solubility indices	
Cellulase dry matter	ED, RI, NVI
Cellulase organic matter	OMD
Cellulose in cupriethylene diamine	DMD, RI, NVI
Digestible laboratory nutrients	DMD
Dry matter	DMD, RI, NVI, ADG
Pepsin dry matter	ED, RI, NVI, DMD, DMI, DDMI
Total after enzymes	DMD
Physical methods	
Fibrousness index	OMD, DMI, DOMI
Particle size index	DMI
In vivo small sample methods	
Nylon bag cellulose digestion	CD
Nylon bag dry matter disappearance followed by acid pepsin	OMD, DMI, DOMI
In vivo artificial rumen dry matter disappearance	DMD
In vitro fermentation methods	
Cellulose disappearance	CD, DMD, RI, NVI, DMI, DDMI
Dry matter disappearance (DMD), one-stage	DMD, DDMI
DMD, with acid-pepsin second stage	DMD, ADG
DMD, with neutral detergent second stage	DMD
Organic matter disappearance	OMD, ADG
Volatile fatty acid production	DMD, ED

† Adapted with permission from Barnes. 1973. Chemistry and Biochemistry of Herbage, Vol. 3, p. 179. Copyright by Academic Press, Inc. (London) Ltd.
‡ ADG = Average daily gain, CD = Cellulose digestibility, CPD = Crude protein digestibility, DDMI = Digestible dry matter intake, DMD = Dry matter digestibility, DMI = Dry matter intake, DOMI = Digestible organic matter intake, ED = Energy digestibility, NVI = Nutritive value index, OMD = Organic matter digestibility, RI = Relative intake, TDN = Total digestible nutrients.

forages have been tested (Table 3.6). No one, simple method is satisfactory for predicting the quality of all forages. Within certain narrow and specified groups of forages, any one of several simple techniques may be acceptable for the prediction of DE content. Prediction of DE intake is much more difficult, however. With many forages, more of the variation in DE intake may be caused by variation in feed intake than in digestibility, and intake and digestibility may not be highly correlated.

The present method of choice for routine forage evaluation is a two-stage in vitro digestion procedure. This involves 1) incubation with rumen microorganisms, and 2) treatment with acid-pepsin. The residue consists of undigested forage cell walls and microbial residues and is very similar to the composition of feces. The in vitro procedure is analogous to the processes of digestion in the animal. This procedure is useful to the researcher, but is not generally available to the forage-livestock manager.

There is no substitute for practical experience when evaluating the quality of harvested forage or pasture. Laboratory analyses, especially for crude protein, a fiber fraction and dry matter, can help greatly, but the wise judgment of the livestock manager is essential. Decisions require knowledge of the animals, especially their size and condition, and a subjective evaluation of the forage, including plant species, maturity, fertilization program, and season when harvested or grazed. If a quick, objective procedure is not available, quality must be evaluated subjectively. Even when an objective forage test is available, forage-livestock managers should not rely completely on it, but should interpret laboratory results in light of their experience.

Factors Affecting Daily Production per Animal

The most important factors affecting daily rate of animal production in a forage-based enterprise are forage factors like quality and availability, and non-forage factors like animal potential and supplemental feeds (Table 3.7). Each will have some effect upon DE intake and efficiency of DE utilization.

Forage Quality

Variations in voluntary DE intake account for major differences in daily production per animal. A number of factors affect intake and digestibility (Table 3.7) and were discussed earlier. The effects of species, maturity, and environment on forage quality will be discussed in Chapter 6.

Forage Availability

If an animal does not have access to as much forage as it would consume voluntarily, its DE intake and rate of production will be less than maximum unless supplemental feeds are supplied. The amount of forage available to each animal is the result of management decisions made by the livestock manager, but confinement and grazing systems differ with respect to the types of decisions to be made.

A. *Confinement.* The amount of hay, silage, or fresh-chopped forage offered daily should be at least 5% greater than consumption if free-choice feeding and maximum (i.e., voluntary) intake are to be

Table 3.7—Factors affecting daily production per animal in a forage-based ruminant production system.

I. Forage factors

A. Quality
   (intake times digestibility)
   1. Composition
      protein
      "fiber"
      minerals
      anti-quality components
      palatability
   2. Physical characteristics
      cell walls
      lignified tissue
      leaf:stem
      degradability
      digestion rate

B. Availability
   (forage/animal/day)
   1. In confinement
      excess offered per feeding
   2. In pasture
      forage yield
      stocking rate
      grazing pressure
      supplemental feed

II. Non-forage factors

A. Animal potential
   (appetite times efficiency)
   1. Animal
      genetics
      age
      sex
      physiological state
   2. Previous treatment
      feeding level
      diseases
      parasites
   3. Environment
      temperature
      humidity
      solar radiation
      precipitation
   4. Biologicals
      feed additives
      implants

B. Supplemental feeds
   (restricted intake)
   1. Improve forage utilization
      protein
      minerals
      vitamins
      additives
   2. Provide digestible energy
      grains
      molasses
      other by-products

achieved. The advantages of insuring maximum intake and providing some opportunity to reject low quality components may more than off-set the amount of waste. Additional effort is involved in weighing the forage offered and refused, but doing so makes it possible to recognize whether animals are eating properly. Removing refused forage from the feed bunks is quite important in hot, humid climates, where mold forms readily.

B. *Grazing.* In general, the same principle of forage availability applies to grazed pastures as to confinement feeding (i.e., some excess of grazable forage should be available). Pasture management is more complex than confinement management because of the dynamic state in a grazed pasture. Forage availability may change rapidly depending on species, fertilization, season, weather, and concentration of animals. If two or more forage species are in association, they may grow at different rates at the same time. Animals grazing the pasture may either increase or decrease pasture yield depending on type and amount of forage and type and concentration of animals. Too many animals on a

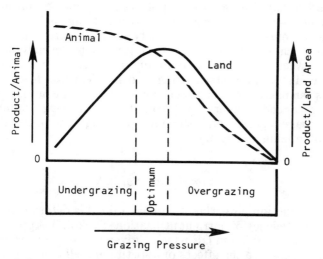

Fig. 3.11—Influence of grazing pressure upon production per animal and production per unit land area. Adapted with permission from G. O. Mott, 1970, Lecture Notes, Agronomy Dep., Univ. of Florida, Gainesville, FL.

pasture may result in decreased yield, as well as decreased forage availability. Conversely, too few animals on a rapidly growing pasture may increase yield and availability, but decrease forage quality because of the accumulation of mature stems. Yields from research plots which are clipped mechanically may not always be applicable to pasture management because clipping techniques do not duplicate effects of animals on forage plants being grazed.

The livestock-pasture manager can exert control on the pasture system by adjusting the number of animals per unit of land area (stocking rate), according to the quantity of forage available per unit of land area, so that there is an optimum amount of forage available per animal. The latter quantity is often expressed as the "grazing pressure" (Fig. 3.11) and is calculated as follows:

$$\text{Grazing pressure} = \frac{\text{forage dry matter/area}}{\text{animal/area}}$$

$$= \text{forage dry matter/animal}$$

Grazing pressure is analogous to the quantity of forage offered daily to animals in confinement. Optimum grazing pressure varies with different forage species, season of the year, and growth stage of the plant, but may be approximately 5% of live body weight. In mixed grass-legume associations, the grazing pressure required to maintain the association may be different from the optimum when one species is used alone.

In practice, it may not be possible to adjust stocking rate as needed to maintain optimum grazing pressure. To do so requires 1) that additional animals be available during periods of rapid pasture growth, and 2) that alternate feed supplies be available for any animals removed from the pasture when availability of forage is low. Management decisions must be compromises between the theoretical optimum and the practical expedient. The stocking rate chosen may be intermediate between maximum and minimum, excess forage may be harvested and stored for later feeding, and supplemental feeds may be provided when pasture production is inadequate for the number of animals being carried. Under range conditions, the large number of different forage species and the low stocking rate intensify the problems of grazing management.

Animal Potential

Daily production potential of a given animal is determined by its appetite and efficiency of nutrient utilization (Table 3.7). Appetite may be defined as the maximum DE which a given animal will consume. It is difficult to separate the effects of appetite and efficiency upon animal potential, because efficiency is related to the level of DE intake above maintenance (Fig. 3.1). For example, some cross-bred steers may be more efficient in converting feed to beef because they consume more DE per day and have a higher daily production. Conversely, if animal potential is increased, greater daily intakes of DE and other nutrients are required in order to achieve the potential. Forage-fed animals will seldom achieve their production potential because, as discussed above, it is seldom possible to achieve maximum daily DE intake with forage as the sole diet. Nevertheless, it is assumed that those factors affecting animal potential (Table 3.7) do influence daily production of forage-fed animals.

The genetic characteristics of the animal are undoubtedly major determinants of animal potential, but the expression of genetic potential may be modified by numerous factors. If a beef steer experiences a period of undernutrition it may, when returned to adequate diet, grow at a rate greater than expected. This "compensatory growth" is a common phenomenon where the feed supply fluctuates during the year. If an animal is parasitized or diseased, or is deficient in some nutrient, it will not achieve its genetic potential. Hormones and other growth-promoting substances may increase animal potential by increasing appetite, efficiency, or both.

Suboptimum environments decrease daily animal production by depressing appetite and/or efficiency of energy utilization. Stress of high heat and high humidity depresses feed intake, decreases grazing time, and reduces production. Some cattle seek shade on hot days, and this behavior may contribute to a lower rate of gain unless they compensate by grazing at night. Environmental stress, animal genetics, and type of diet interact. Selection has produced animals better adapted for specific environments. For example, Zebu or Brahman (*Bos indicus*) cattle were introduced into the southern USA because they were less susceptible to tick infestation and were more heat tolerant than the native cattle.

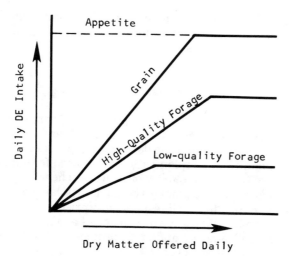

Fig. 3.12—Hypothetical relationships between amount of dry matter offered daily to a ruminant and daily digestible energy (DE) intake, as affected by type of feed and quality of forage.

Appetite, Forage Quality and Forage Availability

Intake of DE, and daily production per animal, are determined by the combined effects of feed availability and feed quality, as well as by animal appetite and potential. Figure 3.12 presents hypothetical relationships between feed availability and feed quality for an animal with a given appetite. The following comments refer to Fig. 3.12. With increasing amounts of feed offered, DE intake increases up to some maximum point for each feed, but that maximum differs among feeds. At all levels of feed offered, DE intake is higher for grain than for forages, and higher for high-quality forage than for low-quality forage. This effect is related to the DE content of these feeds (grain > high-quality forage > low-quality forage) and to intake control mechanisms (Fig. 3.9). Voluntary DE intake equivalent to appetite is achieved only when grain is fed. In practice, the maximum DE intake of very high quality forages, such as alfalfa (*Medicago sativa* L.), may equal appetite—not a frequent occurrence. The level of high-quality forage offered required to achieve voluntary intake is more than the level of grain required to achieve DE intake equal to appetite. Conversely, a low level of forage offered is sufficient to achieve voluntary intake with low-quality forage.

In many forage-livestock systems, forage quality and/or quantity may be limiting and then animal performance is less than the potential. For maintenance of mature animals, low performance is the management objective, and low quality and quantity of forage are sufficient. Where higher performance is the objective, however, managers must decide whether to change the forage program or invest in supplemental feeds. The decision is an economic one, and should be made after careful analysis of the factors limiting DE intake and the cost/return ratios of the alternatives.

Supplemental Feeds

Generally, supplemental feeds are provided to ruminants consuming forage in order to increase the intake of DE and other nutrients. This increase, in turn, will increase daily animal production or decrease weight loss. In addition, supplements may replace forage and extend its use or increase carrying capacity of pasture. The wise and strategic use of supplements may be among the most critical economic factors in the management of a forage-livestock enterprise.

Supplements may 1) improve utilization of available forage by correcting nutrient deficiencies, and/or 2) provide additional DE in the form of grain or by-products (Table 3.7).

A. *Improve Forage Utilization.* Forage utilization may be improved by appropriate nutrient supplements if the forage is deficient in protein, minerals, or vitamins to the extent that forage intake is decreased or animal health and potential are adversely affected.

1. *Protein*—Supplemental protein may increase voluntary forage intake, energy digestibility, and animal performance. The effect of supplemental protein on intake is influenced by forage quality. Mature forages having less than 7% to 8% crude protein are likely to show increased intake due to protein supplementation (Fig. 3.13). Supplemental protein will seldom have a positive effect upon intake of high-quality forage, and may even decrease intake (Fig. 3.13). The animal's protein requirement for maximum production may be higher than that necessary to achieve maximum voluntary DE intake.

If supplemental protein increases intake of low-quality forage, it is unlikely that the level of DE intake achieved will be adequate for much more than maintenance. Mature forage does not have a high potential quality because of a high concentration of cell walls and extensive lignification.

Non-protein N compounds, such as urea, should be used cautiously in supplementing forage. When not properly fed, a large amount of urea can be fatal. Urea may not be well utilized when fed with low-quality forage because of a low concentration of readily available carbohydrates. If the urea is not utilized, the ammonia released from urea in the rumen is absorbed, converted to urea by the liver, and excreted by the kidney. Mixtures of molasses and urea are frequently used as supplements for low-quality forage. Such mixtures provide DE, as well as N, and may also include minerals.

2. *Minerals*—Deficiencies of P, S, Se, Co, Cu, Mg, Mn, I, Zn, Na, and K have been observed in forage-fed ruminants in several locations throughout the world. When these minerals are deficient in the soil and/or forage, dramatic improvements in forage utilization and animal performance may result by providing them in supplements. It is gener-

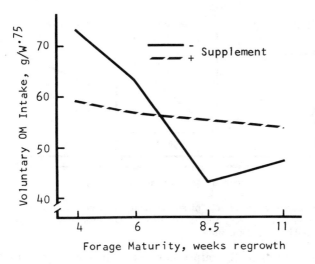

Fig. 3.13—Interaction of forage maturity and supplemental protein on voluntary organic matter (OM) intake by sheep (Pangola digitgrass, *Digitaria decumbens* Stent., hay). Adapted with permission from Ventura et al. (1975).

ally recommended that "trace-mineralized" salt and a P source be available to forage-fed ruminants on a free-choice basis. In some cases, special mineral supplements are required in specific localities during all or part of the year.

3. *Vitamins*—Rumen or intestinal microorganisms generally synthesize adequate quantities of the B-complex vitamins and Vitamin K. Vitamin D is generally synthesized in adequate quantities by solar irradiation of a precursor in the skin of the grazing animal or during sun-curing of hay. Fresh forage and silage will not have Vitamin D, and supplementation may be recommended if such forages are fed to ruminants confined indoors.

Vitamin A supplementation is commonly recommended for all animals. Even though the yellow plant pigment, carotene, is a precursor of Vitamin A, there are reports of improved performance when supplemental Vitamin A is fed to ruminants consuming green, leafy forages and silages. Several inhibitors prevent the conversion of carotene to Vitamin A (for example, a high level of nitrate), but their presence and activity are difficult to predict. The utilization of mature, low-quality forage may be improved by providing Vitamin A supplement.

Alpha-tocopherol, which is the most active form of Vitamin E, normally occurs in forage plants in adequate quantities. Reports of Vitamin E-responsive diseases in ruminants have caused interest in this vitamin and its interrelationship with Se. Weakness and death of calves have been prevented by administering injections of Vitamin E and Se to their dams prior to calving. The occurrence of this deficiency in grazing cattle is difficult to predict.

B. *Provide Additional Digestible Energy.* Supplemental DE in the form of grains or by-products may be used to 1) increase DE intake

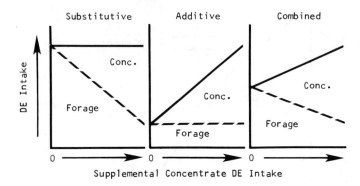

Fig. 3.14—Three hypothetical effects of supplemental concentrate digestible energy (DE) intake upon total (———) and forage (———) DE intake, when supplemental concentrate is fed in limited amounts.

and animal performance, 2) counteract seasonal fluctuations in available forage, 3) increase pasture carrying capacity, or 4) extend the use of harvested forage. Supplemental DE sources will generally be offered in restricted amounts, while forage will be offered free-choice. The amount of supplemental DE an animal will consume can be predetermined, but forage intake could be greatly decreased, or remain the same, depending upon forage quality (Fig. 3.14). With high-quality forages, little increase in total DE intake will be achieved because of substitution. With low-quality forages, forage DE and supplemental DE intakes may be nearly additive. Both substitutive and additive effects may occur simultaneously (Fig. 3.14).

When maintenance or slight weight gain is the objective, low-quality forage alone may be adequate. If forage quality is too low, molasses, citrus pulp, dried poultry waste, or several other commercial byproducts may be used. Substitution of forage DE by supplemental DE may be negligible with low-quality forages, and forage and supplemental DE intakes may be additive. If it is desired to extend the forage, forage intake must be limited by reducing the amount offered per animal.

Growing or lactating animals will generally be consuming higher quality forages for which substitution is expected. Forage DE intake may decrease by 0.5 to 0.8 Mcal for each Mcal of supplemental DE consumed. If the forage is of high quality, it may be more economical to feed forages alone than with supplement and to accept less than maximum daily production per animal. A practical alternative for beef production may be to use forages and grain in separate phases: forages for growing, grain for finishing. Economic conditions and management objectives must be considered in making such decisions.

Animal Production per Unit of Land Area

Animal production per unit of land area is determined by all those factors affecting 1) daily product per animal (quality aspect), and 2) carrying capacity; i.e., animals per unit of land area at optimum grazing pressure (quantity aspect). At a given grazing pressure and forage quality, animal product per unit of land area during the grazing season is related to forage yield. The grazing pressure which maximizes product per unit of land area is different from that which maximizes product per animal (Fig. 3.11).

Supplemental DE may permit a higher stocking rate if there is a substitutive effect of the supplement upon voluntary intake of forage DE (Fig. 3.14). The product per unit of forage-land area will then be increased. Supplementation may be economically advantageous if the cost of feeding the supplement is less than the value of additional product.

SUMMARY

Good managers of forage-livestock enterprises should take into account forage quality, forage availability, animal potential, and supplemental feeds.

Forage quality is best defined as the daily rate of animal production when forage is the only source of energy and protein, the animals have a potential for production, and adequate forage is available. Daily animal production is related to voluntary intake of DE above requirements for maintenance. Major determinants of forage quality are the concentration of cell walls, protein, minerals, and anti-quality components, and the rate of cell wall degradation in the rumen.

Forage availability must be adequate if maximum voluntary DE intake is to occur. Feeding procedures in confinement and adjustments of grazing pressure on pasture determine forage availability.

Animal potential affects forage utilization through effects upon appetite and nutrient utilization. Animal potential is determined by genetic characteristics, age, and physiological state of the animal; previous treatment; nutrient deficiencies; hormones or other additives; environmental stress; and diseases.

Supplemental feeds may increase daily production per animal by improving forage utilization, and by providing additional nutrients, particularly DE. Supplemental DE may be additive or substitutive, depending on forage quality. Supplemental DE may permit higher stocking rates and increase production per unit of land area.

Forage-livestock managers must know the nutrient requirements of animals and the quality and quantity of forage available. Forage quality can vary between wide extremes, and every effort should be made to achieve optimum forage yield, while maintaining adequate quality. Production of high-quality forage, proper feeding or grazing management, and strategic use of supplements are important in determining economic success in forage-livestock enterprises.

SUGGESTED READING

Baker, F., and S. T. Harriss. 1947. The role of the microflora of the alimentary tract of herbivora with special reference to ruminants. No. 2. Microbial digestion in the rumen (and caecum) with special reference to the decomposition of structural cellulose. Nutr. Abstr. Rev. 16:3-12.

Barnes, R. F. 1973. Laboratory methods of evaluating feeding value of herbage. p. 179-214. In G. W. Butler and R. W. Bailey (eds.) Chemistry and biochemistry of herbage. Vol. 3, Academic Press, NY.

Butler, G. W., and R. W. Bailey (eds.). 1973. Chemistry and biochemistry of herbage. Academic Press, NY. 3 vols.

Crampton, E. W., and L. E. Harris. 1969. Applied animal nutrition. 2nd ed. W. H. Freeman and Co., San Francisco, CA. p. 136-164.

————, and L. A. Maynard. 1938. The relation of cellulose and lignin content to the nutritive value of animal feeds. J. Nutr. 15:383-395.

Golding, E. J., J. E. Moore, D. E. Franke, and O. C. Ruelke. 1976. Formulation of hay-grain diets for ruminants. II. Depression in voluntary intake of different quality forages by limited grain in sheep. J. Anim. Sci. 42:717-723.

Harrison, C. M. (ed.). 1968. Forage, economics—quality. Special Publ. 13, Am. Soc. Agron., Madison, WI.

Hungate, R. E. 1966. The rumen and its microbes. Academic Press, NY.

Matches, A. G. (ed.). 1973. Anti-quality components of forages. Special Publ. 4, Crop Sci. Soc. Am., Madison, WI.

McDonald, I. W. 1968. The nutrition of grazing ruminants. Nutr. Abstr. Rev. 38:381-400.

Montgomery, M. J., and B. R. Baumgardt. 1965. Regulation of food intake in ruminants. 1. Pelleted rations varying in energy concentration. J. Dairy Sci. 48:569-574.

Mott, G. O. 1973. Evaluating forage production. p. 126-135. In M. E. Heath, D. S. Metcalfe, and R. F. Barnes (eds.) Forages. The Iowa State Univ. Press, Ames.

National Research Council. 1971. Nutrient requirements of domestic animals. No. 3. Nutrient requirements of dairy cattle. National Academy of Sciences, Washington, DC.

————. 1975. Nutrient requirements of domestic animals. No. 5. Nutrient requirements of sheep. National Academy of Sciences, Washington, DC.

————. 1976. Nutrient requirements of domestic animals. No. 4. Nutrient requirements of beef cattle. National Academy of Sciences, Washington, DC.

Raymond, W. F. 1969. The nutritive value of forage crops. Adv. Agron. 21:1–108. Am. Soc. Agron., Madison, WI.

Van Keuren, R. W. (ed.). 1974. Systems analysis in forage crops production and utilization. Special Publ. 6, Crop Sci. Soc. Am., Madison, WI.

Van Soest, P. J. 1967. Development of a comprehensive system of feed analysis and its application to forages. J. Anim. Sci. 26:119–128.

Ventura, M., J. E. Moore, O. C. Ruelke, and D. E. Franke. 1975. Effect of maturity and protein supplementation on voluntary intake and nutrient digestibility of Pangola digitgrass hays. J. Anim. Sci. 40:769–774.

Wilkinson, J. M., and J. C. Taylor. 1973. Beef production from grassland. Butterworths, London.

Part II. Genetic and Environmental Effect on Quality

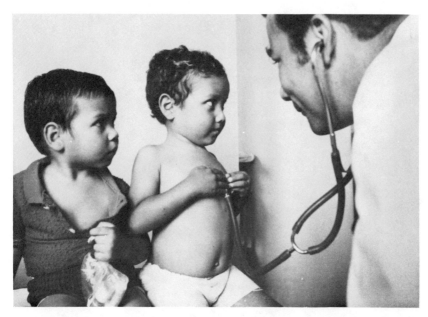

Chapter 4

Proteins in Food and Feed grain Crops

L. S. BATES AND E. G. HEYNE

Kansas State University
Manhattan, Kansas

Proteins are prime foodstuffs. They, along with carbohydrates and fats (the "fuelstuffs"), provide the majority of our daily food. Because proteins have a limited active life, there is a continual demand for protein in the diet. Thus man and all other animals consume, break down, resynthesize, and excrete protein and protein metabolic products each day. In fact, the old adage "we are what we eat" may be true in the sense that the proteins we eat are broken down to their component amino acids which are subsequently incorporated into enzymes, muscles, blood albumins, and a multitude of other proteins for orderly cellular growth and body maintenance.

The more closely the ingested proteins meet body needs, the better is the nutritional quality of the food or feed for that particular organism. Thus for man's needs, most meat proteins are nutritionally superior to plant proteins. Although some nations are protein-rich relative to animal or other good-quality proteins, most countries are not, unfortunately; and a general shortage of all protein resources is recognized worldwide. In many countries, population pressure and cultural taboos long ago caused shifts along the food chain from animal protein to plants—the primary producers of protein. In protein-rich countries a total shift to plant proteins will be long coming for many reasons. However, it is logical to assume that some shifting of resources worldwide will require all of us to produce and use more plant proteins in the future. Consequently, there is an immediate need for information— basic biochemical genetics and applied plant breeding—to redirect the priorities for our food and feed grain crops.

Table 4.1—Typical cereal, legume, and oilseed structural comparisons.

Cereal	Legume	Oilseed
Zea mays L. Maize or corn	*Glycine max* Merr. Soybean	*Linum usitatissimum* L. Linseed or flax
Pericarp†	Pod†	Capsule†

Seed
1) seed coat
2) endosperm (major
   storage tissue)
	Seed
	1) seed coat
	2) no endosperm (nutrient
	   storage in cotyledons)
		Seed
		1) seed coat
		2) endosperm (a few cell
		   layers; most nutrient
		   storage in cotyledons)

3) embryo
   scutellum (small
   cotyledon)
   embryonic axis
	3) embryo
	   two cotyledons
	   embryonic axis
		3) embryo
		   two cotyledons
		   embryonic axis

† Pericarp, pod, and capsule are terms for the remaining ovary wall within which the seed(s) develop. In maize and other monocots, the pericarp is not separated from the seed and together they make up the whole fruit or caryopsis.

PLANT PROTEIN AND SEED STRUCTURE

Plant proteins come in various packages—seeds, leaves, roots or tubers, fruits, saps or exudates, and other forms. Seeds, those independent capsules with reserve nutrients and the capacity to reproduce plant life, have been the most useful to man throughout the ages. Seeds are individually wrapped and protected for easy harvesting and handling. Their metabolism is low when dry, and thus they can be stored readily for long periods of time. Most importantly, the proteins from several seed sources may complement each other so that they nearly meet the needs of man and his domestic animals for growth and body maintenance. An excellent example of complementation is the classic corn (*Zea mays* L.) and bean (*Phaseolus vulgaris* L.) diet that has nourished the population of Latin America for centuries. Roots and tubers also feed large populations, and leaves, fruits, saps, and other protein sources supplement the diet. However, they are all secondary to seed proteins in overall importance to world nutrition.

Monocotyledonous and Dicotyledonous Protein

Seed-producing crop species, with which we are concerned here, are subdivided into two classes—monocotyledonous and dicotyledonous. Cereal grains and other grasses are monocotyledonous; all other seed crops—pseudocereals (traditionally utilized in the same manner as cereals), legumes, and oilseeds—are dicotyledonous. The distinction is important because endosperm/cotyledon relationships help one to visualize seed protein quality in general.

Endosperm, the primary storage tissue of monocotyledons, is composed of starch and lesser quantities of protein and other nutrients. It is an excellent food and energy source because the thin-walled cellular structure is readily ruptured and the nutrients made available.

Table 4.2—Average % composition of food and feed grains.

Grain	Protein	Carbohydrates	Fats
Barley	13.1	76.0	2.1
Beans, dry	22.9	57.2	1.4
Bean, mung (*Vigna radiata* L. Wilczek)	23.9	56.9	1.1
Buckwheat (*Fagopyrum esculentum* Moench.)	12.6	71.6	2.9
Chickpea	20.1	54.0	4.3
Corn, dent	10.4	81.2	4.5
Cottonseed	23.1	26.3	22.9
Cowpea (*Vigna sinensis* L. Endl.)	23.4	55.8	1.3
Flaxseed	24.0	24.0	35.9
Millet, proso	13.1	71.4	4.0
Oat groats	16.2	63.7	6.1
Peanut, without hulls	30.4	11.7	47.7
Rice, brown	9.4	86.8	1.8
Rye	13.4	80.1	1.8
Safflower	16.3	17.5	29.8
Sesame	22.3	10.9	42.9
Sorghum	12.5	79.2	3.4
Soybean	37.9	24.5	18.0
Sunflower, without hulls	27.7	16.3	41.4
Wheat, avg. all types	14.2	79.1	2.0

Cotyledons, fleshy storage organs of embryos, contain predominantly protein, oil, and some carbohydrates but essentially no starch. The cells are also thin-walled and the nutrients readily available. In monocotyledons, the scutellum is the single cotyledon with a chemical composition similar to that of the dicotyledons (Table 4.1).

The relative proportion of endosperm to cotyledon varies between and within species. In cereals, the contribution of the scutellum is small in comparison to the endosperm, which makes up roughly 85% of the total seed. In dicotyledons the reverse is generally true. Endosperm tissue is one or only a few cell layers thick. Because the most abundant nutrient in endosperm is starch, one would expect cereals to be lower in protein than either legumes or oilseeds but just as high in metabolic energy. Table 4.2 illustrates typical seed compositions.

Proteins and Protein Quality

Although no seed proteins from a single plant species are nutritionally balanced for humans or other monogastric animals, some proteins can be isolated from seeds which have an amino acid balance almost identical to meat proteins. That is, the range of proteins in either plants or animals is large and one can find numerous similarities. Unfortunately, the cereal storage proteins that predominate in the endosperm are of low biological value and hence lower the total protein quality of seed. Some cotyledon proteins are similarly deficient with respect to one or more amino acids and consequently lower the biological value of seeds, depending on their relative concentrations.

The measure of biological value commonly used is protein efficiency ratio (PER) calculated as the weight gain/protein intake for any growing animal. The PER reflects the quality of a food or feed as a measure of the availability of individual amino acids. Because histidine, isoleucine, leucine, lysine, methionine, phenylalanine, threonine, tryptophan, and valine cannot be synthesized by mammals, they are essential in the diet. Essentiality varies with age. Histidine is required for children but probably not for adults. Moreover, if all the amino acids are not available at each protein synthesis site, 1) synthesis will stop, 2) the partially completed protein will be recycled for its constituent amino acids, and 3) much energy will be wasted by the organism essentially "spinning its wheels" without gain.

The first limiting amino acid, that one essential amino acid in lowest concentration relative to specific animal requirements which limits protein synthesis, affects PER and, of course, growth and maintenance of the organism. Because amino acids occur primarily in proteins, more information about the types of proteins and their location within seeds is needed for an understanding of protein quality and the genetic and environmental factors affecting it.

Protein Classification and Distribution

Proteins are generally classified according to their solubility in various aqueous solvents. It is an old and practical system with some overlapping between groups. The major groups and criteria are:

albumins — soluble in salt-free water
globulins — insoluble in salt-free water but soluble in dilute neutral salt solutions
prolamins — insoluble in water or neutral salt solutions but soluble in 60–80% ethyl alcohol or in dilute alkali
glutelins — insoluble in water, neutral salt solutions, or alcohol but soluble in dilute alkali and acids

Albumins and globulins serve numerous regulatory, reserve, transport, enzymatic, and other essential functions. They are synthesized early in seed development to assure germination and survival of the species. In some species they are the reserve nutrient proteins. In general they are comparable to good animal proteins and are well balanced nutritionally for food uses.

Prolamins and glutelins are primarily endosperm storage proteins that are synthesized somewhat late in seed development and serve as reserve proteins for seedling vigor in the subsequent generation once the embryo has germinated. For our nutritional needs, glutelins are less well balanced than albumins and globulins, but are far superior to prolamins in this respect. Prolamins, of the four major protein groups, are nutritionally inferior because they are practically devoid of lysine and tryptophan—two of the essential amino acids.

The relative proportions of the four protein groups in seeds determine the overall nutritional quality. When albumins and globulins predominate, quality is good. If large amounts of prolamins are synthesized at the expense of other proteins, particularly albumins and globulins, the nutritional quality of seeds diminishes. If additional pro-

Table 4.3—Prolamin content of selected cereals.

Cereal	Prolamin
	% of seed protein
Corn, normal	47
Corn, opaque-2	23
Corn, floury-2	22
Corn, opaque-2/floury-2	15
Corn, Illinois High Protein	73
Corn, Illinois High Protein opaque-2	44
Barley, normal	38
Barley, hiproly	27
Sorghum	60
Italian millet (Setaria italica L. Beauv.)	50
Bread wheat (Triticum aestivum L. em Thell.)	45
Durum wheat (Triticum turgidum L.)	60
Rye	40
Oats	12
Rice	8

lamins are synthesized as the total seed protein content is increased, the nutritional value may be increased somewhat but at a lowered efficiency of N utilization.

The relative distribution of various seed proteins determines what post-harvest processes may be used, the potential utilization of the products, and existing limits for genetic manipulations. In monocotyledons, albumins and globulins are found predominantly in the embryo and its related nurse tissue, the scutellum, and in the aleurone (outermost cell layer of the endosperm). Scutellum contains about 30% of high nutritional-quality protein but represents less than 2% of the seed of wheat (Triticum spp.) and other small grains and up to about 10% in the larger grained cereals. Similarly, aleurone represents less than 10% of the total seed and has limited impact on protein quality. Of course small amounts of albumins and globulins exist in every endosperm cell as part of the metabolic machinery, but storage proteins predominate. There is also a decreasing protein-concentration gradient from the outer endosperm cells immediately below the aleurone layer to the central endosperm cells. The endosperm becomes increasingly soft and starchy toward the center of the kernel. Because endosperm is approximately 85% of the seed and contains 70 to 80% of the total protein, cereal breeding for protein-quality improvement centers around endosperm-modifying genes. Some breeding for increased germ size can increase protein quality but only within the limits of maximum scutellum size.

Two cereals, oats (Avena sativa L.) and rice (Oryza sativa L.) are exceptional for their very low prolamin contents (Table 4.3). Their storage protein synthesis, obviously quite different from that of all other cereals, may hold clues for improving all cereal protein quality because the protein is well-balanced nutritionally. Rice and some oat cultivars exhibit protein quantity problems, however, and are being bred intensively for higher protein. Relatively high protein cultivars of several species are now available.

In soybean (*Glycine max* L. Merr.), peanut (*Arachis hypogaea* L.), dry beans (*Phaseolus* spp.), peas (*Pisum* spp.), and other dicotyledons, little true endosperm exists in the mature seed. Instead, nutrient reserves are located in the two cotyledons. The proteins are similar to the well-balanced proteins of the corresponding monocotyledon and consist of albumins, globulins, and glutelins. Prolamins are essentially nonexistent. Thus, dicotyledon seed protein quality is understandably superior nutritionally to cereal protein quality; one needs only to compare the relative amounts of cotyledons or endosperm to the total seed to visualize the differences.

We have not escaped our seed protein-quality dilemma yet, however. Despite the superior nutritional rating of dicotyledon protein, it is still low in methionine—another essential amino acid—and subsequently not a complete food or feed. Fortunately, monocotyledons and dicotyledons complement each other and amino acid deficiencies are balanced as mentioned earlier. A cereal-legume diet occasionally supplemented with small amounts of meat, dairy products, fish, vegetables, insects, or other protein sources is reasonably well balanced. Other dicotyledons, particularly oilseeds, also produce excellent protein sources for varying a predominantly cereal-based diet. Inexpensive and available dicotyledonous sources of high protein include soybean, peanut, dry beans, peas, sunflower (*Helianthus annuus* L.), safflower (*Carthamus tinctorius* L.), flax (*Linum usitatissimum* L.), sesame (*Sesamum indicum* L.), cotton (*Gossypium* sp.), crambe (*Crambe abyssinica* Hochst. ex. R. E. Fries), various mustards (*Brassica* sp.), and many more. Those can supplement the cereals wheat, rice, corn, barley (*Hordeum vulgare* L.), sorghum (*Sorghum bicolor* L. Moench), rye (*Secale cereale* L.), oats, millets (*Pennisetum* spp., *Panicum miliaceum* L., and *Eleusine coracana* L. Gaertn), and triticale (*X Triticosecale* Wittmack) which have lower protein contents. Because proteins are digested at different rates, the greater the number of different complementary proteins one ingests, theoretically the more superior and sustaining the diet becomes for cellular growth and maintenance.

GENETIC EFFECTS ON PROTEIN QUANTITY AND QUALITY

Approximately 63% of the world's total protein resources are consumed by man directly as grains or grain products, according to many estimates. If one considers the amount of grain-based feeds consumed by domestic animals which is subsequently eaten by man as meat or dairy products, the estimates would go even higher. Even so, many people in developing nations consume in excess of 80% of their dietary protein as grains. Consequently, plant-improvement specialists throughout the world are working to enhance the nutritional quality and the total quantity of food and feed grain resources.

Protein Quantity and Yield

Early protein/yield studies of cereals showed that as protein content increased, yield decreased. The phenomenon was observed also in oilseeds in which selection for oil yield decreased protein content. Given that only so much energy can be fixed photosynthetically by plants, it is logical that additional protein synthesis must compete with the synthesis of other nutrients for the limited energy available.

Thus, protein/yield or protein/oil yield trade-offs can be understood in terms of balancing relative energetic costs as calculated by Bhatia and Rabson. A more detailed discussion of energy follows in chapter 5.

In order to improve both protein content and yield simultaneously, one would expect increased photosynthetic or physiological efficiency from a cultivar, given limited energy. Although it cannot be classified as an efficiency breakthrough without further study, one wheat cultivar in recent years has indicated the protein/yield barrier in cereals may be manipulated. 'Plainsman V,' a little heralded wheat from western Kansas, has produced yields up to 96% of the best check cultivar and at the same time had 2 to 3% more protein than standard cultivars. Similar advances can be expected in other wheat cultivars and in other crops as breeding priorities shift more toward protein quantity and quality improvements.

Protein Quality and Yield

Protein quality improvements traditionally have been considered more difficult than protein quantity improvements because of the ease with which germplasm could be composited and selected to increase total protein. The classic example is 'Illinois High Protein' corn, which has been selected for more than 70 generations. Total protein is now approximately 27% of the kernel. The number of genes involved is unknown but the genes are suspected to be primarily those controlling prolamin and glutelin synthesis and their modifiers because of the high concentration of those two protein groups.

In 1963 to 1965, the effects of the *opaque-2* and *floury-2* genes on protein synthesis in corn were discovered and reported. These genes, in essence, reversed the quantities of prolamin and glutelin synthesized in the endosperm and resulted in a net two-fold increase of lysine and tryptophan. The nutritional effect of *opaque-2* was even more dramatic —monogastric animals gained 3.5 (average) to 12 times faster than on normal corn (Fig. 4.1), and children with advanced cases of protein malnutrition (kwashiorkor) recovered completely (Fig. 4.2). The potential impact of *opaque-2* corn on world nutrition stimulated scientists and administrators to explore other cereals and field crops for protein quality genes.

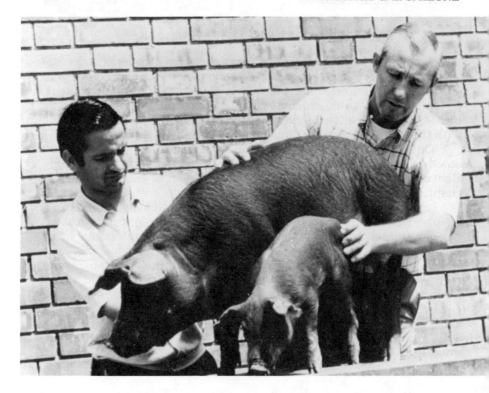

Fig. 4.1—Swine fed either opaque-2 (larger animal) or normal (smaller animal) corn from 35 to 165 days of age. Photograph courtesy of the Rockefeller Foundation. Described by Maner (1975).

Screening for protein nutritional quality in diploid crops produced quick results. Discoveries of high lysine genotypes were made in barley ('Hiproly' and 'Risø 1508') and sorghum (high lysine). The *opaque-2*, Risø 1508, and high lysine sorghum genes are all classed as opaque types and have similar effects on protein synthesis. In addition to a reversal of prolamin and glutelin synthesis in the earliest developed material, more albumins and globulins were formed, the seeds became opaque to visible light, yield was lowered somewhat, and stored-grain insects and microorganisms became greater problems.

It is difficult to envision all these secondary problems as effects of a single gene with current gene models. In addition, the physiology of genotypes that affect protein quality are not understood. However, the primary effect of the recessive *opaque-2* gene is to reduce the proportion of prolamin (zein) protein. This reduction may occur as a function of ribonuclease degradation of zein mRNA. If that is so, zein bodies or protein bodies, the granular storage form of zein in corn, should be absent or mostly so in *opaque-2* maize; and indeed they are. Zein-glutelin ratio changes, the lack of protein granules, and an apparently different packing arrangement of starch in the protein matrix of

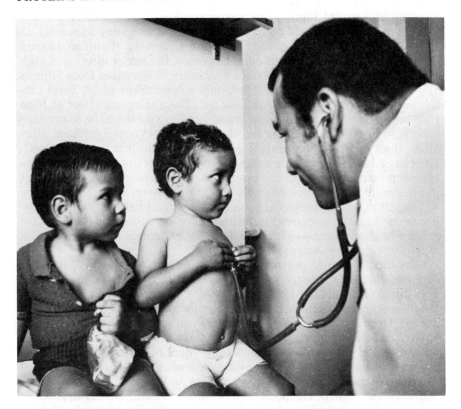

Fig. 4.2—Dr. A. G. Pradilla examining two 3-year-old boys who made medical history by recovering completely from seemingly fatal kwashiorkor using opaque-2 corn as the only source of protein in their diet. The disease is characterized by protein malnutrition producing retarded growth, lack of appetite, diarrhea, edema, abnormal hair color and texture, and eventually death. Photograph courtesy of the Rockefeller Foundation.

the endosperm are believed to alter light diffraction patterns and consequently produce an opaque appearance of the kernel. The packing arrangement may be responsible also for part of the yield reduction encountered in some *opaque-2* hybrids. However, some *opaque-2* hybrids yield well, perhaps better than the normal counterparts, and many investigators have reported modifiers which produce near normal crystalline endosperms. Not all secondary problems have been solved yet, but they undoubtedly will be as plant breeders work through new germplasm.

In polyploid cereals no opaque-type genes have been discovered. This lack may be due to 1) a masking effect of proteins synthesized by the additional genome(s), or 2) a low natural frequency of "opaque" possibly common ancestry, and because the mutant Risø 1508 has already been found in barley, one can anticipate future discoveries of high lysine genes in wheat and its wild relatives and presumably also in rye.

The absence of high lysine genes in wheat has stimulated a few groups around the world to hybridize wheat and barley expressly to improve the nutritional quality without adversely changing baking quality because of lowered gliadin. However, the major effect of a high lysine gene should be on prolamin synthesis. Segregates from Illinois High Protein corn × *opaque-2* lost only a few tenths of 1% total protein, whereas lysine percentage doubled. The analogous effect of Risø 1508 on high protein bread wheat would not be expected to be so dramatic because of genome buffering but the boost should be a significant milestone if an opaque-type control gene from barley functions in wheat. Effects of lower gliadin on baking remain conjectural until the genotype can be produced and the bread baked. However, it is known that gliadin controls loaf volume (gas retention power) whereas glutenin, mixing time. Minor technical problems for bread baking would probably occur but the improved nutritional quality would be an asset for other wheat products—cakes, cookies, pastas, gruels—which do not require strong gas retention properties.

In dicotyledons, particularly legumes, an analogous "opaque" type would undoubtedly be a high methionine protein segregate. Screening for high methionine-containing legumes and protein synthesis studies are underway, but no significant increases have been reported. Because increases of cellular free amino acids are inconsistent with maintenance of pH balance for orderly growth and development, and with current concepts of genetic control of protein biosynthesis, amino acid balance improvement must be considered only in terms of the relative proportion of various proteins and their constituent amino acids. Consequently the problems of improved amino acid balance in legumes are quite unlike those of improving cereals. Instead of breeding to limit one low quality protein type—prolamins—one must seek genes or gene combinations to produce new globulins or perhaps new albumins with more of the sulfur-containing amino acids, methionine and cystine. It is difficult to conceive of producing those recombinant types. Additionally, the subsequent proteins would undoubtedly exhibit considerable secondary structural and solubility changes and would no longer be globulin but perhaps glutelin instead. An exaggeration perhaps, but unfortunately one would not expect to obtain those extensive protein synthesis rearrangements, however desirable. There are simply no clear-cut answers to breeding improved cotyledon protein at the moment.

High methionine-containing oilseeds exist, so genes are available for synthesizing dicotyledon protein(s) that are considerably different from those normally produced. Sesame, of all oilseeds, stands out as an excellent source of high methionine protein. Other sesame breeding problems have, however, minimized protein quality studies in this species. Instead, oilseed breeding has emphasized yield, and that yield is measured by the primary product—oil, fiber, etc. When protein is a definite byproduct of fiber or oil or when protein is lowered in quality during processing, little gain can be expected in protein quality. In soybean, protein is equally as important as, or more important than, the oil. Additionally, the protein products (particularly concentrates and isolates) lack only a few tenths of 1% methionine of being a nutritionally balanced protein and may encourage, at least in soybean, a major effort to increase the methionine level.

A related consideration in protein-quality breeding is the control of natural plant toxins. The toxic alkaloid gossypol has been bred out of the cotton seed with no loss in cotton quality (presuming adequate chemical insect control) and with a great increase in usable protein with minimal processing costs. Breeding to lower levels of toxins in chickpea (*Cicer arietinum* L.), broadbean (*Vicia faba* L.), and other legumes would improve the utilization of these excellent protein sources.

New Techniques

Interspecific and intergeneric hybridization (wide hybridization) offer many new possibilities for modifying protein quality and quantity. One of the immediate methods to expand the primary gene pool of a species and enhance recombination is to increase the use of secondary and tertiary gene pool germplasm. Some wide hybrids may serve as bridging species to transfer advantageous genes between genera or may become new crop species themselves, like triticale.

The F_1 of many interspecific and most intergeneric hybrids are sterile. The full chromosome complement of both parental species must be doubled or the hybrids maintained vegetatively as clones. In some intergeneric crosses, a second type of hybridity exists in which only a few genes are introgressed from the paternal into the maternal chromosomes. This latter process holds great promise for the transfer of specific genes if the mechanism(s) can be controlled by the breeder.

Somatic cell fusion is another route to wide hybridization. Naked protoplasts, obtained by enzymatically dissolving the cell wall from leaf or other vegetative cells, are fused under low-speed centrifugation in the presence of polyethylene glycol. Very wide hybrids—monocotyledon × dicotyledon—can be made, but it is exceedingly difficult to regenerate hybrid plants from a single hybrid cell. Regeneration of complete hybrid plants from fused, interspecific hybrid cells has been accomplished in a few cases and somatic fusion techniques are continually improving. Somatic cell fusion may play an important role in future protein breeding.

ENVIRONMENTAL EFFECTS ON PROTEIN QUANTITY AND QUALITY

Protein quality and quantity, particularly quantity, are influenced by environment and heredity. By definition, the phenotype is the visible manifestation of interactions between genotype and environment. A superior cultivar reflects the fitness of that cultivar to a particular environment. When a cultivar is found to have wide adaptability to different climates and environmental fluctuations, one can presume that the genotype codes for metabolic machinery with wider temperature, pH, and/or other parameter optimums and subsequently the cultivar can respond and yield favorably within a wider range of environmental stresses than can nonadapted cultivars. We have little control over environmental factors except for irrigation and fertilizer.

Temperature and Water Stress

Although several books, reviews, and journal articles in recent years have explored temperature and water relationships in plants, virtually nothing is known about the effects of these factors on protein quantity and quality. Even less is known about the relative rates of protein synthesis and degradation under suboptimal growth conditions. In general, protein synthesis in plants proceeds most rapidly at temperatures which produce highest overall growth rates. However, protein content can increase or decrease at either temperature extreme.

Water deficits also cause reduced protein synthesis accompanied by an accumulation of free amino acids. The increased amino acid levels are believed due to slowed protein synthesis and not to protein degradation.

Nitrate reductase, an unstable enzyme whose synthesis reflects environmental conditions, is a key control point in nitrate utilization. It may be a major factor in temperature/water stress effects on protein synthesis also. Other enzymes for N utilization and enzymes for protein synthesis for which environmental stress data are limited precludes a full understanding of protein synthesis/environmental stress. Additional complications arise because albumins, globulins, prolamins, and glutelins are synthesized in the seed at different times in different species. It is impossible, at present, to determine the extent and nature of particular stresses until the seed, the final product, is harvested. Even then, the events of stresses cannot be deciphered.

In general, the effect of either temperature extreme or of water stress, is to increase the protein quantity and quality of seed protein on a percentage basis. Plants tend to channel nutrients into developing seeds rather than to sporophytic or vegetative tissue, thus improving chances of viable seed production. Because embryo proteins are primarily albumins and globulins of good nutritional quality and prolamins are synthesized later in endosperm development, stress affected plants tend to have more favorable amino acid balances than plants grown under more favorable conditions. Also, proportionally more proteins than oil or storage carbohydrates are synthesized in the early stages of seed development, producing an "improvement" of protein quality and quantity in a stress-damaged seed. That kind of "improvement" is worthless because of lowered yield and generally shriveled seed.

Other Stresses

Insect damage, wind shredding of leaves, disease, and all other factors that stress a plant at seed-filling time can be expected to influence protein deposition in the seed. In plants as in humans, the nutritional requirements must change dramatically during stress situations, requiring a mobilization of reserves and energy to the site(s) of damage or infection. The effect on seed protein biosynthesis is indirect but definite because of metabolite drain from seed production. As with temperature and water stress, the particular effect on seed-protein quality and quantity depends on the stage of seed development.

Nitrogen Fertilizer Effects

We are still limited to a few cereal genotypes—those containing "opaque-type" genes and rice and oats—which can respond to ideal conditions of fertilizer and water and not lose protein nutritional quality. Application of the principles discussed in the *Protein Classification and Distribution* and *Protein Quality and Yield* sections of this chapter would lead one to that conclusion. The relationships have been experimentally demonstrated. Extensive work has been done on cereal protein quantity/N fertilizer relationships. The protein content of a cultivar may increase, decrease, or not respond appreciably to N fertilizer depending on genotype, existing available soil N supplies, environmental stresses, type of fertilizer (ammonia, urea, nitrate, or other form), and time and method of application. In general, cereal protein content increases proportionally to N applied. Late foliar urea applications are particularly effective in increasing total protein. In experiments with 'Pawnee' wheat, protein increased from 9.3 to 16.1% in one particular year. Generally, the increase was in the range of 2 to 4% (Fig. 4.3). More recent work demonstrated that late N applications increased protein percentage without decreasing yield.

In legumes, there is some question whether symbiotic N-fixing bacteria meet total plant N needs for growth as reported for soybean. Because grain legumes do not generally yield more in response to N fertilizer applications, it has been assumed that N-fixation supplies needs. However, failure to respond positively to N-fertilizer applications is better explained as a plant (or plant/bacteria) preference for already fixed N. Nitrogen-fixing activity decreases proportionately to available fixed N giving no net gain in total N input or output. Some organic N sources more compatible with the soybean N-fixing system have increased yields 20%. Because the total theoretical energy cost to the legume is the same for reducing N to ammonia or reducing nitrate to ammonia and the limiting factor for legume N-fixing in the field is photosynthate, one might conclude it best to maximize all environmental conditions possible for legumes except N and then select cultivars for increased yield presuming that increased protein will be ob-

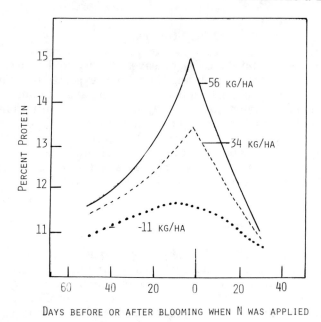

Fig. 4.3—Effects of foliar N applications before and after flowering on seed protein content of Pawnee wheat. From Finney (1957).

tained too.

The rest of the dicotyledons require N fertilizer the same as cereals to obtain an adequate yield of total dry matter and protein. Because cotyledon protein is well-balanced nutritionally, it pays to increase N applications to the optimum for any crop species assuming the cultivars are efficient depositors of seed protein and there is a market available to justify the expense of extra protein synthesis.

In addition to N fertilizer, nontoxic levels of simazine and terbacil (herbicides) have been shown to increase seed-protein levels. Although herbicide action is related to low N levels and the mechanism(s) of these compounds in low concentration is unknown, it is suspected they directly affect nitrate reductase enzymes and subsequently N assimilation. One might speculate that in the future, more specific protein-biosynthesis regulators will become available for large-scale aerial spraying. Currently, however, the economic returns are not favorable.

SUMMARY

In summary: 1) yield is inversely proportional to protein content except for a few recently developed cultivars; 2) protein quantity is inversely proportional to protein quality except for "opaque" genotypes in cereals, but dicotyledonous plants do not exhibit such limitations; and 3) once the essential proteins for reproduction have been synthesized, excess N fertilization increases yield of total grain and protein per unit land area but lowers protein quality per unit protein in cereals. Because this effect is essentially an endosperm phenomenon, dicotyledons do not exhibit that limitation.

Most importantly, because protein quality breeding—despite its importance—is still in its infancy, we can expect tremendous improvements in high quality protein synthesis both in understanding basic events and in practical application of basic biochemical genetics toward improving the world's protein resources.

SUGGESTED READING

Alexander, D. E., J. W. Dudley, and R. J. Lambert. 1971. The modification of protein quality of maize by breeding. Eucarpia 5:33–43.

Altschul, A. M. 1958. Processed plant protein foodstuffs. Academic Press, NY.

Bates, L. S. 1966. Protein and amino acid composition of opaque-2 maize endosperm. Part I. M.S. Thesis, Purdue Univ., Lafayette, IN.

————, K. A. Mujeeb, and R. F. Waters. 1976. Wheat × barley hybrids: Problems and potentials. Cereal Res. Commun. 4:377–386.

Bauman, L. F., E. T. Mertz, A. Carballo, and E. W. Sprague (eds.). 1975. High-quality protein maize. Dowden, Hutchinson and Ross, Inc., Stroudsburg, PA.

Bhatia, C. R., and R. Rabson. 1976. Bioenergetic considerations in cereal breeding for protein improvement. Science 194:1418–1421.

Bliss, F. A., and T. C. Hall. 1977. Food legumes—compositional and nutritional changes induced by breeding. Cereal Foods World 22:106–113.

Boyer, J. S. 1976. Stress relationships in protein synthesis. p. 159–171. In Genetic improvement of seed protein. Proc. of a workshop, Washington, D.C., 18–20 Mar. 1974, National Academy of sciences, Washington, DC.

Bushuk, W. (ed.). 1976. Rye: production, chemistry, and technology. American Association of Cereal Chemists, St. Paul, MN.

Eppendorfer, W. H. 1975. Effects of fertilizers on quality and nutritional value of grain protein. p. 249–263. In Proc. 11th Colloq., International Potash Institute, Bornholm, Denmark.

Fajersson, F. 1961. Nitrogen fertilization and wheat quality. Agric. Hort. Genet. 19:1–195.

Finney, K. F., J. W. Meyer, F. W. Smith, and H. C. Fryer. 1957. Effect of foliar spraying on Pawnee wheat with urea solutions on yield, protein content, and protein quality. Agron. J. 49:341–347.

Hageman, R. H., R. J. Lambert, D. Loussaert, M. Dalling, and L. A. Klepper. 1976. Nitrate and nitrate reductase as factors limiting protein synthesis. p. 103–134. In Genetic improvement of seed proteins, Proc. of a workshop, Washington, D.C., 18–20 Mar. 1974, National Academy of Sciences, Washington, DC.

Hallsworth, E. G. (ed.). 1958. Nutrition of the legumes. Academic Press, NY.

Hardy, R. W. F., and U. D. Havelka. 1975. Nitrogen fixation research: A key to world food. Science 188:633–643.

Harlan, J. R. 1975. Crops and man. Am. Soc. Agron., Madison, WI.

Havelka, U. D., and R. W. F. Hardy. 1974. Legume N_2 fixation as a problem in carbon nutrition. Proc. Int. Symp. Nitrogen Fixation, Washington State Univ. Press, Pullman, WA.

Houston, D. F. (ed.). 1972. Rice: chemistry and technology. Am. Assoc. Cereal Chemists, St. Paul, MN.

Hulse, J. H., and E. M. Laing. 1974. Nutritive value of triticale protein. International Development Research Center, Ottawa, Canada.

Inglett, G. E. (ed.). 1972. Seed proteins. AVI Publ. Co., Inc., Westport, CT.

Ingversen, J., B. Kóie, and H. Doll. 1973. Induced seed protein mutant in barley. Experientia 29:1151.

Jimenez, J. R. 1966. Protein fractionation studies of high lysine corn. E. T. Mertz and O. E. Nelson (eds.) In Proc. high lysine corn conference. Corn Refiners Assoc., Inc., Washington, DC.

Johnson, V. A., P. J. Mattern, J. W. Schmidt, and J. E. Stroike. 1973. Genetic advances in wheat protein quantity and composition. p. 547–556. In E. R. Sears (ed.) Fourth int. wheat genetics symp. Univ. of Missouri, Columbia, MO.

Jones, J. G. W. (ed.). 1973. Biological efficiency of protein production. University Press, Cambridge.

Konzak, C. F. 1977. Genetic control of the content, amino acid composition, and processing properties of proteins in wheat. Adv. Genet. 19:407–582.

Maner, J. H. 1975. Quality protein maize in swine nutrition. p. 58–82. In L. F. Bauman, E. T. Mertz, A. Carballo, and E. W. Sprague (eds.) High-quality protein maize. Dowden, Hutchinson and Ross, Inc., Stroudsburg, PA.

Mertz, E. T., L. S. Bates, and O. E. Nelson. 1964. Mutant gene that changes protein composition and increases lysine content of maize endosperm. Science 145:279–280.

Miezan, K., E. G. Heyne, and K. F. Finney. 1977. Genetic and environmental effects on the grain protein content in wheat. Crop Sci. 17:591–593.

Millerd, A. 1975. Biochemistry of legume seed proteins. Ann. Rev. Plant Physiol. 26:53–72.

Milner, M. (ed.). 1975. Nutritional improvement of food legumes by breeding. Wiley and Sons, NY

Morrison, F. B. 1959. Feeds and feeding. The Morrison Publishing Co., Clinton, IA.

Mosse, J. 1966. Alcohol soluble proteins of cereal grains. Fed. Proc. 25:1663.

Munck, L. 1972. Improvement of nutritional value in cereals. Hereditas 72:1–128.

——–, K. E. Karlson, A. Hagberg, and B. O. Eggum. 1970. Gene for improved nutritional value in barley seed protein. Science 168:985–987.

Nelson, O. E. 1969. Genetic modification of protein quality in plants. Adv. Agron. 21:171–194.

——–. 1973. Breeding for specific amino acids. p. 303–311. In A. M. Srb (ed.) Genes, enzymes, and populations. Plenum Press, NY.

——–, E. T. Mertz, and L. S. Bates. 1965. Second mutant gene affecting the amino acid pattern of maize endosperm proteins. Science 150:1469–1470.

Pomeranz, Y. 1971. Composition and functionality of wheat-flour components. p. 585–674. In Y. Pomeranz (ed.) Wheat: chemistry and technology. Am. Assoc. of Cereal Chemist, St. Paul, MN.

Recommended Dietary Allowances. 1974. 8th Ed. National Academy of Sciences, Washington, DC.

Rendig, V. V., and D. S. Mikkelsen. 1976. Plant protein composition as influenced by environment and cultural practices. p. 84–106. *In* B. H. Beard and M. D. Miller (eds.) Opportunities to improve protein quality and quantity for human food. Publ. 3058. Univ. of California, Davis, CA.

Ries, S. K. 1968. Spray on protein boosters. Crops Soils 20(8):15–17.

Siman, G. 1972. Nitrogen status in growing cereals. Royal Agricultural College of Sweden, Uppsala.

Singh, R., and J. E. Axtell. 1973. High lysine mutant gene (hl) that improves protein quality and biological value of grain sorghum. Crop Sci. 13:535–539.

Smartt, J. 1976. Tropical pulses. Longmans, London.

Wassom, C. E., and R. C. Hoseney. 1973. Selection of high lysine corn using a scanning electron microscope. Crop Sci. 13:462–463.

Wolf, M. J., U. Khoo, and H. L. Seckinger. 1967. Subcellular structure of endosperm protein in high-lysine and normal corn. Science 157:556–557.

Chapter 5

Carbohydrates and Lipids in Food and Feed Grain Crops

L. M. GOURLEY AND R. G. CREECH

Mississippi State University
Mississippi State, Mississippi

Plants, by means of the photosynthetic process, trap some of the sun's radiant energy and concentrate this energy in specialized storage organs. Carbohydrates, proteins, and oil or fats are the primary high-energy compounds utilized by man and animals in their nutrition. Proteins are discussed in chapter 4. In this chapter we will discuss the quantity and quality of these high-energy compounds, and genetic and environmental factors affecting energy in food and feed grain crops.

CARBOHYDRATES

Carbohydrates are readily classified into three general groups. The simple sugars like glucose or fructose are called monosaccharides. When two or more, but fewer than 10 simple sugars are bonded together like sucrose and raffinose they are called oligosaccharides. The tetrasaccharide stachyose contains many combinations of sugars found in plants and seeds (Fig. 5.1). Those carbohydrates, composed of a larger number of simple sugar molecules, such as starch and cellulose, are called polysaccharides. Cellulose is the most abundant of all organic compounds.

Of the oligosaccharides, the disaccharide sucrose is the most important. Sucrose occurs throughout the plant kingdom in seeds, leaves, fruits, flowers, and roots. It is synthesized primarily in the leaves and serves as the transport medium to the various parts of the plant. Sucrose and its phosphorylated derivatives function as the energy source for the metabolic processes and as a carbon source for biosynthesis of the various cellular components.

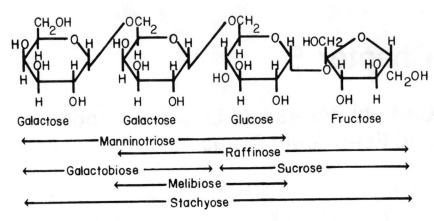

Fig. 5.1—Molecular structure of stachyose and possible component sugars formed by degradative enzymes.

Starch is a mixture of glucans (polymers made up of only glucose molecules) and is utilized as the principal food reserve polysaccharide and also as an energy reserve by the plant during growth. Starch is the chief source of carbohydrate in the human diet. After cellulose, starch is probably the most commercially utilized of all the polysaccharides. The two components, amylose (chains of glucose) and amylopectin (chains with branches), represent the major portion of the starch material (Fig. 5.2).

Starch is found mainly in the plant kingdom where it is laid down in the form of granules, which are insoluble in the cell medium. It occurs in several parts of the plant—leaves, stems, shoots, and storage organs such as tubers, rhizomes, and seeds.

Starch granules are formed in the chloroplasts when any excess carbohydrate is available from photosynthesis. As an energy need develops in buds, roots, or immature storage organs, assimilated starch is converted to sucrose and transported through the phloem to all living cells. In the leaves of monocotyledonous cereal plants, starch is scarcely detectable. In this case, sucrose is the primary reserve carbohydrate in the leaves and it is transported to other specific organs and converted to starch.

Starch Sources and Uses

The principal industrial starch of the western hemisphere is recovered from corn (*Zea mays* L.). Starch comprises about 70% of the total dry weight of mature corn kernels and is composed of approximately 25% amylose and 75% amylopectin. Although the USA is the largest producer of cornstarch, only a small percentage of the corn crop is used for this purpose. About one-half of the milled starch is used for making derived products such as sugars, syrups, fructose syrups, and dextrins.

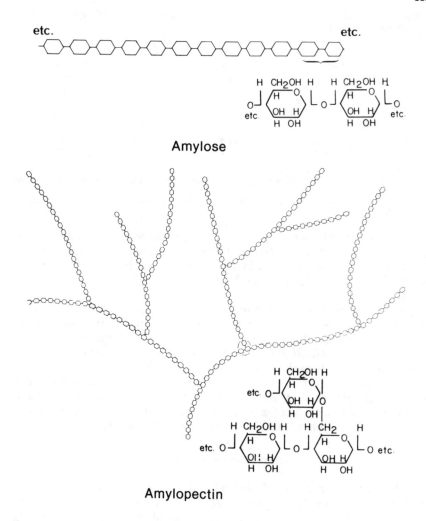

Fig. 5.2—Molecular structures of amylose and amylopectin.

   Cornstarch is an important manufactured product because of its many industrial applications. Watson describes the preparation and industrial utilization of cornstarch. Unmodified cornstarch is used extensively in paper and paper product manufacture, glues and adhesives, and is also the raw material from which modified starches and corn syrups are made. Oxidized starch has the properties desired for paper and textile sizing, and laundry finishing. Precooked foods, bakery and prepared home mixes, and convenience foods require pregelatinized starches which form pastes in cold or warm water.

   The starch of the waxy endosperm mutants, 99 to 100% amylopectin starch, develops a reddish purple coloration when stained with an iodine solution. Iodine induces a deep blue color in high amylose or

Table 5.1—Content of total sugars and photosynthetic efficiency of sugarcane plants during development.†

| | | Total sugars (fresh wt.) | | |
Time‡	Day/night temperatures	Top 100 cm (39.4 in) of stalk	Remainder of stalk	Photosynthetic efficiency§
days	C F	———————%———————		rate
0	24/19 (75/66)	6.8	11.1	1.15
35	17/10 (63/50)	8.9	11.8	0.69
55	17/10 (63/50)	10.4	12.8	0.44
81	24/19 (75/66)	10.4	12.8	0.76
121	24/19 (75/66)	11.7	13.9	0.80

† Adapted from Waldron et al. (1967).
‡ Six-month-old plants were equilibrated at 24/19 C (75/66 F) for 1 month prior to the commencement of measurements.
§ No. of grams of carbohydrate fixed per calorie of visible light radiation $\times 10^5$ = photosynthetic efficiency.

normal cornstarch. Amylopectin starches are especially suitable as gels for thickening in canned and frozen foods and, when modified, for many other specialized uses. High amylose starch has been used for edible clear films and moisture-resistant coatings for food and other industrial products.

Starch is also isolated from several other plant sources. Potato (*Solanum tuberosum* L.) starch occupies an important position in European markets. Tapioca starch, which comes from the tuberous roots of cassava (*Manihot esculenta* Crantz), is an important source of starch in Asia. Starch is obtained from the roots of arrowroot (*Maranta arundinacea* L.). Sago starch is derived from the pith of various palms. Starch can also be isolated and purified from grain sorghum [*Sorghum bicolor* (L.) Moench], cereals, sweet potatoes (*Ipomoea batatas* Lam.), and several other plant species.

Sugar Sources and Uses

The primary sources of commercial sucrose are sugarcane (*Saccharum officinarum* L.), sugarbeet (*Beta vulgaris* L.), sugar sorghum, and the sap of maple trees (*Acer saccharum* Marsh). The sugarbeet is the principal source of sucrose in temperate countries of the world. Maple syrup and sugar are commercially produced in the USA, primarily in the New England states. Sweet sorghums grow well either in the tropical or temperate zones, but sugarcane is limited to the tropical and subtropical production areas of the world.

Sucrose constitutes about 50% of the total dry matter in a mature stalk of sugarcane. The sugar content of mature internodal tissue without rind (mainly storage parenchyma cells) may attain 80% of the dry weight. Internodes from the top portion of the stalk contain much lower levels of sucrose (Table 5.1) and may contain glucose and fructose at levels of about 5% fresh weight. Using ^{14}C labeled sucrose, it has been demonstrated that sucrose can move through the phloem system of one stalk into other tillers within a plant without breakdown and resynthesis. The vacuole of the storage parenchyma cells appears

Table 5.2—Quantities of various carbohydrates and total dry matter in 28-day-old kernels of several maize genotypes.†

Genotype	Reducing sugar	Sucrose	Total sugar	WSP‡	Starch	Total carbo- hydrates	Dry matter
				%			
normal	0.8	2.2	3.0	2.2	73.4	78.6	43.8
ae	1.9	7.4	9.4	4.4	49.3	62.9	37.5
du	1.3	6.7	8.0	1.9	59.9	69.8	38.9
sh2	3.6	22.1	25.7	5.1	21.9	52.8	26.3
su1	3.9	4.4	8.3	24.2	35.4	69.6	37.6
su2	1.4	1.9	3.3	1.9	64.6	69.8	43.6
wx	1.6	1.7	3.3	2.2	69.0	74.5	37.3
ae su1	2.1	5.3	7.4	3.2	34.4	45.1	33.9
ae wx	3.2	12.3	15.4	4.6	39.5	59.5	28.3
du wx	3.0	9.5	12.5	11.6	45.4	69.5	34.8
sh2 su1	5.7	20.1	24.5	4.9	15.7	45.4	24.6
du su1 wx	2.3	6.7	9.0	47.5	15.9	72.3	35.3
ae du wx	4.4	23.7	28.1	4.9	32.0	65.1	24.5

† Adapted from Creech (1968).
‡ Water-soluble polysaccharides.

to be the main storage compartment, since it is the only component large enough to contain the high sucrose levels attained in mature storage tissue.

In sweet sorghum cultivars used for syrup, it is desirable to have rapid sucrose inversion. Inversion is the process in which sucrose is hydrolyzed by acids or enzymes to a mixture of equal amounts of fructose and glucose. The mixture is called invert sugar. In cultivars used for sugar production a low rate of sucrose inversion is required. Small quantities of starch and aconitic acid are troublesome in both syrup and sugar production from sorghums. Commercial production of sugar from sorghum is being accomplished with slight modifications of processing equipment used for sugarcane.

GENETIC EFFECTS ON CARBOHYDRATES

Corn is especially appropriate for the study of the effects of genetic interactions on carbohydrate metabolism because of the availability of many gene mutations that influence carbohydrate properties. Data for these gene systems can apply, in some instances, to sorghum, rice (*Oryza sativa* L.), barley (*Hordeum vulgare* L.), and other cereals. Vavilov's Law of Homologous Series, in general terms, states that if mutant genes affecting the characteristic of interest are available in other similar plant species they can be found or established in the species under investigation. Because of the importance of corn to industry and to human and animal nutrition, the effect of specific genes and gene combinations on carbohydrate metabolism has been extensively investigated.

Table 5.2 contains data on the quantity and quality of carbohydrates in 28-day-old kernels of some single, double, and triple mutant genotypes of corn. In general, all of the mutant genotypes have significantly less starch than the normal genotype. The normal dent corn kernel contains about 72% starch by dry weight and free sugars range between 1 to 3%, with sucrose as the major component. The mean gross energy value of corn is around 4,000 cal/g. Starch contributes about 80% of this total.

One of the early mutants reported as being different from normal dent corn in appearance and carbohydrate composition was waxy (wx). This gene produces an endosperm containing mostly amylopectin starch and only a trace of amylose. High-amylopectin starch permits maximum water absorption and susceptibility to enzymatic hydrolysis. The resultant variation in grain digestibility indicates that nutritive efficiency can be altered. Also, waxy genes have been found in sorghum, rice, and barley.

A wide range in starch granule digestibility is conditioned by waxy and other corn genotypes. A recessive gene, amylose-extender (ae), gives the starch a high resistance to enzymatic digestion. The ae endosperm produces starch with about 60% amylose. This specialty starch has found many industrial applications. High-amylose genetic variants of peas (Pisum sativum L.) and barley have been identified.

The gene sugary-1 (su1), found in standard sweetcorn hybrids, causes an accumulation of water soluble polysaccharides (WSP) which are principally composed of phytoglycogen, a highly branched polyglucan similar to animal glycogen. Shrunken-2 (sh2) produces a dramatic increase in sucrose content without an increase in WSP. The sh2 mutant is epistatic (dominance of one gene over a non-allelic gene) to su1 resulting in a decreased accumulation of WSP and an increase in sucrose content (Laughnan, 1953). The mutant gene dull (du) causes a slight increase in sucrose but the triple mutant genotype du su1 wx nearly doubles the accumulation of WSP. Several of these mutants, singularly and in combination (ie. sh2, su1 sh2, and ae du wx), are used to improve the quality of fresh market sweetcorn.

Genetic variability of sugarcane stalk and sucrose yield does exist and is probably quantitatively inherited. Because sugarcane is polyploid, specific genes for quality or quantity of carbohydrates are difficult to identify.

In sorghum, the characteristic for sweet juice is recessive to the allele for non-sweet juice. A single dominant gene is responsible for the dry-stalk (low juice content) condition. Both of these genes influence the accumulation of sugars in sorghum stalks used for sugar or syrup production.

In summary, cultivars of most crop species differ for carbohydrate quantity and quality and there is genetic variability for these characteristics. Whenever a component of a storage tissue is altered, it is usually at the expense of another component. Many carbohydrate mutant genotypes of corn produce less starch and frequently less total crop yield. By genetic manipulation, seed and plant carbohydrate quality can be changed to enhance human and animal biological efficiency. Unless a price differential for improved quality is paid, however, farmers and marketing specialists still consider the main attribute of a new plant cultivar to be yield. When more information about metabolic regulation and gene-enzyme or enzyme component relationships are discovered in corn, this information may be applied to many plant species.

ENVIRONMENTAL EFFECTS ON CARBOHYDRATES

Nearly every environmental change will have some effect on a growing plant. Such environmental factors as photoperiod, light, temperature, water, and fertility affect growth and reproduction in all food and feed grain crops. The specific effect of an environmental change on quantity and quality of carbohydrates depends upon plant stage of growth and the duration of the change. An environmental shock in the vegetative stage of growth could be rather inconsequential to crop production; however, in the reproductive stage an environmental change of only a few days duration could be highly detrimental.

Because sugarcane is vegetatively propagated, and therefore the plants are genetically identical, the environmental effects on carbohydrate accumulation in this crop have been studied extensively. Sugarcane is an exceptionally efficient plant and, when grown in the tropics, can utilize 6 to 7% of the sunlight available for photosynthesis (Woodwell, 1970). Sugarcane, corn, sorghum, and several other important food plants are C-4 plants (the initial product of photosynthesis is a 4-carbon acid), while most cereal and temperate plants are C-3 plants. For these reasons, and the fact that sucrose is the carbohydrate storage reserve in sugarcane, this species is ideal for environmental studies.

Under controlled conditions, sugarcane produces the greatest stalk diameter and the most total sugar per plant at constant day and night temperatures of from 22 to 30 C (72 to 86 F) under a 12-hour photoperiod (Grasziou et al., 1965). The constant-temperature effect is much reduced when the photoperiod is decreased to 8 hours. This constant temperature effect is in contrast with that of many temperate plant species which show a general growth requirement for thermoperiodicity. In these studies with sugarcane, the environments that gave the highest rate of dry matter production also yielded the most sugar per plant and the highest sugar content on a dry weight basis (12%).

Sugar contents may vary from 12 to 20% of fresh weight in the field. Peak values for sucrose concentration in sugarcane juice occur at 18° north or south latitude (Shaw, 1953). This peak appears to be due to an optimum combination of seasonal temperature and daylength variations. At latitudes closer to the equator the temperature variation is too small for both optimum growth and sucrose storage. At higher latitudes the daylengths during the winter period are too short.

Producers have observed that apparently moderate drought stress reduces vegetative growth and increases sugar content expressed as percentage of dry weight in sugarcane. C. E. Hart (1967) has

Table 5.3—Starch content, granular diameter, and amylose content of the developing potato tuber.†

Size of tuber‡		Starch content (fresh wt.)	Avg. granular diameter§	Amylose#
cm	(in)	%	μ	%
0–1	(0–0.4)	5.0	10.5	12.5
1–2	(0.4–0.8)	5.6	N.D.	13.2
2–3	(0.8–1.2)	6.4	20	13.9
3–4	(1.2–1.6)	8.3	N.D.	14.6
4–5	(1.6–2.0)	9.2	25	16.4
5–6	(2.0–2.4)	11.0	28	17.2
6–7	(2.4–2.8)	13.4	32	18.2
7–8	(2.8–3.1)	17.0	N.D.	18.5
8–9	(3.1–3.5)	17.5	43	19.0
9–10	(3.5–3.9)	17.5	N.D.	19.2
10–11	(3.9–4.3)	18.0	45	19.8

† Adapted from Geddes et al. (1965).
‡ Longitudinal diameter.
§ N.D. = not determined. ($\mu = 3.937 \times 10^{-5}$ in).
# Calculated from ({ iodine affinity ÷ 19.5} × 100).

suggested that the process of translocation to growing points is slowed in plants with low moisture supply, thereby giving more time for the transfer of sucrose to storage. Growth (and presumably respiration) is reduced, providing an excess of sugar for storage. The opposite response has been observed in sugar sorghum. abundant moisture will produce a luxuriant growth of the sugar sorghum plant with a lower sugar percentage and frequently a lower total yield of sugar.

Temperature shocks have been known to increase the sucrose content of leaves. Long-term treatments at temperatures below 20 C (68 F) affect the partitioning of photosynthate between sugar storage and stalk elongation in sugarcane. Low temperatures were particularly favorable for sugar storage and generally in high accumulations of sucrose in leaf tissue. However, photosynthetic efficiency decreases fairly rapidly at day temperatures of 17 C (63 F) (Table 5.1).

Plant diseases cause a decrease in carbohydrate accumulation. Stem rust greatly decreases sugar content, mostly the sucrose fraction, in seedling wheat leaves (Krog et al., 1961). Reduction in sucrose content is closely related to severity and kind of rust infection. The sucrose reduction in infected leaves is also more pronounced in a 29 C (84 F) environment than in one of 18 C (64 F). As in sugarcane, warm temperature favored sucrose translocation, in this case to the rust fungus, while the cool temperature promoted sucrose storage and thus less reduction by the rust infection.

Implications on the storage or translocation of sucrose can be summarized as follows:

1. Sugarcane exhibits more rapid growth and produces more sugar per plant at constant warm temperature under a 12-hour photoperiod than under other environmental regimes. Optimum growth and sucrose storage occurs at 18° north or south latitudes, however, due to the combination of seasonal temperature and daylength variations.

2. Dry weather reduces translocation and plant growth and increases sucrose expressed as percentage of dry weight in sugarcane.

Table 5.4—Content of starch, amylose, sucrose, and reducing sugars during corn endosperm development.†

Kernel age	Starch Per endosperm	Dry wt.	Amylose in starch	Sucrose Per endosperm	Dry wt.	Reducing sugars Per endosperm	Dry wt.
days	mg‡	%	%	mg‡	%	mg‡	%
8	0.01	1.2	9.0	0.02	2.0	0.29	38.2
10	0.02	1.5	9.0	0.08	8.6	0.45	35.0
12	0.36	9.2	8.5	0.75	19.2	1.05	26.8
14	7.63	46.0	18.0	2.16	13.0	1.50	9.1
16	21.51	62.2	20.8	3.30	9.6	1.82	5.3
22	60.55	76.9	27.5	2.97	3.8	1.14	1.5
28	92.01	77.0	26.5	1.85	1.6	0.65	0.6

† From Tsai et al. (1970).
‡ (mg = 1.543×10^{-2} grains or 3.527×10^{-5} oz).

3. Abundant moisture produces a lush growth and low sucrose percentage in sugar sorghum.

4. Low temperature reduces growth and favors sugar storage in the leaves of sugarcane and wheat.

5. Leaf diseases reduce plant growth and sucrose content by diverting photosynthate for use by the pathogen.

Environmental changes also affect starch quantity and quality. As the potato tuber matures, starch content, granule size, and amylose increase. Table 5.3 shows these effects of tubers harvested at 10-day intervals. Similar changes seem to occur in the developing corn kernel.

Net accumulation of reserve starch in the developing corn endosperm commences approximately 10 to 12 days after pollination (Table 5.4). Starch accumulation increases rapidly, both on a per endosperm basis and as a percentage of dry weight, from 8 to 28 days after pollination. Reducing sugars per endosperm, on the other hand, increase up to the 16-day stage and then decline, but when expressed as a percentage of dry weight, they decline during the entire period. Sucrose content peaks at 12 days, then decreases. Amylose, expressed as a percentage of starch, increases rapidly from 12 to 14 days and thereafter more slowly until the 22-day stage, when maximal values are attained. These results are due to specific enzyme patterns and indicate that an environmental change from 12 to 14 days after pollination could significantly alter starch biosynthesis.

The effects of location and years of experimentation on amylose synthesis in corn endosperm were studied by V. L. Fergason and M. S. Zuber (1962). Corn cultivars differing in amylose content were grown at eight locations in the USA for a 3-year period. The highest average amylose content was obtained at a Wisconsin location that had the lowest cumulative degree days (cumulative degrees of temperature for days with average temperature above 10 C or 50 F), while the lowest amylose content was obtained from a North Carolina location that had the second highest cumulative degree days. The authors suggested that the influence of low temperature on high amylose synthesis should be considered by farmers growing high-amylose corn hybrids for commercial production.

In general, for most cereals and starchy tuberous species, starch yield is directly associated with total yield. As the storage organ matures, whether fruit or seed, the percentage of amylose in the starch increases. Any environmental stress during biosynthesis of storage starch would affect total starch and crop yield, and the amylose:amylopectin ratio. Low cumulative degree days favor a higher-than-normal percentage of amylose in corn starch.

Improving the Biological Value of Carbohydrates

Discussion of carbohydrates will be limited here to the overall improvement of biological value for human food and livestock feed, including the structural modification of a polysaccharide genetically or by environment to make it more digestible. Reducing or eliminating some anti-quality factors will also be discussed in this section.

Amylose:Amylopectin Ratio

The amylose:amylopectin ratio in any starch is controlled primarily by genetic factors. Plant breeders are attempting to develop new crop cultivars varying in amylose and amylopectin contents for specific industrial and food and feed purposes without substantial decreases from normal yields. Three species of interest with more than 50% amylose are genetic variants of peas, corn, and barley. Waxy (high amylopectin) cultivars of corn, rice, sorghum, and barley are known and are currently under investigation.

Other factors beside the amylose:amylopectin ratio must be considered when comparing rice quality. Nutritional studies have been conducted on 'Hamsa,' a rice cultivar with a higher than normal amylose starch, and 'IR-8,' a high yielding rice cultivar with a lower than normal amylose starch (Rao, 1971). In rice, cooking quality and palatability are associated with amylose content. A low amylose content rice is classed as a sticky cultivar of poor cooking quality by USA standards. It was observed that IR-8 rice was more slowly digested by alphaamylase than Hamsa, and it produced a lower blood glucose level than the Hamsa cultivar in humans. Thus the solubility and digestibility of starch granules are complex, depending not only upon molecular differences but also upon the association of these molecules.

Maltose in Sweet Potatoes

Carbohydrate quality can be improved by controlled enzymatic hydrolysis of starch. Maltose content greatly influences the culinary quality and processed products made from sweet potatoes. Normal maltose content of cured, stored sweet potatoes now can be obtained in a few minutes compared to the several weeks formerly required. By temperature activation of beta-amylase, the primary saccharifying enzyme in sweet potatoes, maltose can be increased to 38% of the dry weight of freshly harvested ground sweet potato (Hoover and Harmon, 1967). No significant changes occur in other sugars or total carbohydrate content, and the treatment allows the processing of sweet potatoes directly from the field.

Raffinose and Stachyose

Minor sugars in small quantities can act as anti-quality factors. Two oligosaccharides found in plants in small quantities, raffinose and stachyose, have been identified as causing flatus (gastrointestinal gas) in humans and other animals. Edible beans, including defatted soybean meal, cause the common complaints of nausea, cramps, intestinal pain, diarrhea, and the egestion of rectal gas. Individuals vary widely in the amount of flatus produced, probably due to the microflora spectrum in the intestinal tract (Rackis et al., 1970).

Plant breeders have attempted, unsuccessfully to date, to find strains of beans without raffinose and stachyose. European workers have used mutation breeding in an attempt to block the synthesis of enzymes responsible for these two minor carbohydrates. With more and more soybean flour being used to extend meat products in human and pet foods, these anti-quality factors will become more of a problem.

Fructose and Xylose

The relative sweetness and other unique properties of sugars can be important in human nutrition and health. Due to recent advances in technology, starch can be converted to high-fructose syrups on an industrial scale. Manufactured high-fructose corn syrup (HFCS) is being substituted for sucrose and corn syrup dextrins in candy, ice cream, and other confections. From its beginning in 1972, HFCS production in the USA was estimated to be 1.1 billion kg (2.4 billion lb) for 1976 (Andres, 1976). Fructose is sweeter than sucrose, requiring fewer calories to satisfy the nation's sweet tooth in what is becoming a health problem—over-weight Americans.

New "sugarless" gum is sweetened by xylitol, the sugar alcohol made from xylose the five-carbon sugar. Xylitol equals sucrose in sweetness and energy value. Finnish scientists have reported that xylitol does not cause tooth decay and does not require insulin in metabolism. Some agricultural residues, for example cornstalks, contain 15 to 30% xylose and could become commercial sources of xylitol.

LIPIDS—FATS AND OILS

The simple lipids are fatty acid esters of various alcohols. In oilseeds the most abundant of the simple lipids are the triglycerides, which are fatty acid esters of glycerol. If the three fatty acids of the

Table 5.5—Representative fatty acids found in triglycerides of fats and oils.

Fatty acid	Formula
	Unsaturated acids
Oleic	$HOOC (CH_2)_7 CH = CH (CH_2)_7 CH_3$
Linoleic	$HOOC (CH_2)_7 CH = CH CH_2 CH = CH (CH_2)_4 CH_3$
Linolenic	$HOOC (CH_2)_7 CH = CH CH_2 CH = CH CH_2 CH = CH CH_2 CH_3$
Erucic	$HOOC (CH_2)_{11} CH = CH (CH_2)_7 CH_3$
Ricinoleic	$HOOC (CH_2)_7 CH = CH CH_2 CHOH (CH_2)_5 CH_3$
	Saturated acids
Caprylic	$HOOC (CH_2)_6 CH_3$
Capric	$HOOC (CH_2)_8 CH_3$
Lauric	$HOOC (CH_2)_{10} CH_3$
Myristic	$HOOC (CH_2)_{12} CH_3$
Palmitic	$HOOC (CH_2)_{14} CH_3$
Stearic	$HOOC (CH_2)_{16} CH_3$
Arachidic	$HOOC (CH_2)_{18} CH_3$

triglycerides are saturated they will be solid at room temperature and are called fats.

$$
\begin{array}{cccc}
CH_2OH & \text{Fatty acids} & CH_2O\text{--}OR_1 & CH_2O\text{--}OR_1 \\
| & + & | & | \\
HCOH & \text{enzymes} & HCO\text{--}OR_1 \quad \text{or} & HCO\text{--}OR_2 \\
| & \xrightarrow{} & | & | \\
CH_2OH & & CH_2O\text{--}OR_1 & CH_2O\text{--}OR_1 \\
& (-H_2O) & & \\
\text{Glycerol} & & \text{Simple} & \text{Mixed} \\
& & \text{triglyceride} & \text{triglyceride}
\end{array}
$$

(See Table 5.5 for examples of fatty acids). If the constituent fatty acids are unsaturated, they will be liquid at room temperature and are called oils. Thus, melting point reflects the degree of saturation of the fatty acids. Compound lipids are composed of different compounds in addition to fatty acids.

In higher plants, reserve fatty acids accumulate mainly as triglycerides in the seed or fruit. They act as food stores for the embryo, aiding it to successfully produce the next generation of the species. Oils and fats contain more than twice the quantity of energy, gram per gram, than do carbohydrates. Other structures of higher plants also contain fatty acids, although almost always in smaller quantities. In leaves, stems, and roots, they tend to be combined as glycolipids and phospholipids. These polar lipids are associated with metabolic and structural functions. Fruit-coat fats and seed oils provide a concentrated source of lipids and their extraction and sale has produced a huge commercial industry throughout the world.

The major fatty acids referred to in this chapter are all saturated and unsaturated monocarboxylic (one COOH) acids with a straight even-numbered carbon chain and are usually found in nature as triglycerides. The even-numbered carbon chain in fatty acids and the specific location of double bonds in the unsaturated acids are effected by specific enzymes in biosynthesis. The saturated and unsaturated fatty acids shown in Table 5.5 all commonly occur in higher plants. These 12 fatty acids account for over 90% of those in the world's commercial vegetable fats and oils (Hitchcock and Nichols, 1971). Unsaturated fatty acids are usually found more frequently and abundantly than saturated fatty acids.

The quantity and quality of oil among genera of several plant families are listed in Table 5.6. This table is a partial listing of some of the species of plants used for edible, non-edible, and pharmaceutical oil around the world. The quality and quantity of oil will vary depending on the method of extraction, the cultivar used, and area in which the plant is grown. Table 1.13 in Chapter 1 lists world oil production data for several plant species.

Different structures of the plant frequently vary in the fatty acid composition of the oils. For example, the major fatty acids from the pericarp (palm oil) of the palm fruit (*Elaeis guineensis* Jacq.) are palmitic, oleic, and linoleic acids, while the oil from the nut of the same fruit (palm kernel oil) consists chiefly of lauric, myristic, palmitic, and oleic acids.

Fatty Acid Composition of Major Crops

For years the more popular salad and cooking oils in the USA have been corn, peanut (*Arachis hypogaea* L.), cottonseed (*Gossypium hirsutum* L.), and to a lesser extent olive (*Olea europea* L.), sunflower (*Helianthus annuus* L.), and safflower (*Carthamus tinctorius* L.). From the proportion of fatty acids shown in Table 5.6, the polyunsaturated/saturates (P/S) ratio of these popular salad and cooking oils can be calculated. They all have P/S ratios of about 4.0 to 9.0 except cottonseed oil which has about 2.5. All have linoleic acid contents of 45% or more except olive and peanut oils which have about 80 and 60% oleic acid, respectively. Keeping quality is good because they all have 1% or less linolenic acid.

Corn oil is essentially a byproduct of the wet and dry milling industry. The corn kernel contains about 5% oil, 85% of which is in the germ. The high price of corn oil has assured that all of the oil goes into food products. Peanuts and cotton are not grown primarily for their oil, although peanut oil usually commands a premium price in the salad and cooking oil markets. Crushing peanuts to produce edible oil is of secondary importance to the direct edible use of peanuts and peanut products, which consumes approximately two-thirds of the USA production.

Because of the demand for vegetable oil and the wide geographical range of soybean [*Glycine max* (L.) Merrill] production, soybean oil has become the most important domestic vegetable food oil. Its non-food uses have declined to less than 10% of the food uses. This drop is due, in part, to its relatively low cost, which is possible because of the high value of the protein meal. The soybean contains about one part oil for two parts protein. This ratio in other oilseeds is usually reversed.

Table 5.6—Quantity of oil and approximate fatty acid composition of some typical grains and other vegetable sources.

Scientific name	Common name	% oil	Fatty acid composition (%)												
			Unsaturated acids					Saturated acids							
			18:1 Oleic	18:2 Linoleic	18:3 Linolenic	22:1 Erucic	18:1 Ricinoleic	8:0 Caprylic	10:0 Capric	12:0 Lauric	14:0 Myristic	16:0 Palmitic	18:0 Stearic	20:0 Arachidic	Other
Gramineae:															
Zea mays L.	corn	3–6	40	45	1							9	3		
Oryza sativa L.	rice (bran)	12–18	45	35								15	2		
Triticum aestivum L.	wheat	1–2	25	49	21										
Avena sativa L.	oats	4–10	58	31								10			
Sorghum bicolor L. Moench	grain sorghum	3–5	38	47	1							8	5		
Palmae Cocoineae:															
Elaeis guineensis Jacq.	oil palm (pulp)	64–88	46	6							2	39	5		
	oil palm (kernel)	46–53	15	1				3	5	49	16	8	2		
Cocos nucifera L.	coconut (copra)	62–71	7	2				8	10	48	15	6	2		
Cruciferae:															
Brassica napus L.	rape	39–45	27	18	4	50									
Leguminosae:															
Arachis hypogaea L.	peanut	43–48	59	20								9	3	6	
Glycine max L. Merrill	soybean	16–20	27	54	6							9	4		
Linaceae:															
Linum usitatissimum L.	flax	36–40	20	21	51										
Euphorbiaceae:															
Ricinus communis L.	castor bean	46–50	6	2			85					3	3		7
Malvaceae:															
Gossypium hirsutum L.	cotton	20	25	46								22	5		
Oleaceae:															
Olea europea L.	olive	10–60	79	8								11			
Pedaliaceae:															
Sesamum indicum L.	sesame	48–54	44	42								8	4		
Compositae:															
Helianthus annuus L.	sunflower	40–52	33	57											9
Carthamus tinctorius L.	safflower	25–30	24	65	1										9

Table 5.7—Flavor or odor intensity values of soybean oils in heat and room odor tests at different levels of linolenate.†

Flavor or odor‡	% Linolenate			
	0.0	1.3	2.0	7.8
Flavor				
Rancid	1.2	1.2	1.4	1.4
Painty	0.3	0.7	1.0	1.6
Fishy	0.0	0.1	0.0	0.8
Odor				
Hot soil	0.7	--	--	0.2
Rancid	0.1	--	--	0.6
Fishy	0.2	--	--	1.4

† From Cowan et al. (1970).
‡ Flavor or odor intensity value = (weak responses + 2 × medium responses + 3 × strong responses) divided by the number of tasters.

The use of soybean oil in food was unacceptable for a period of time because its unsaturation and high content of linolenic acid (6% or more) created a flavor and odor stability problem (Table 5.7). Unprocessed soybean oil lacks sufficient linolenic acid for paints and possesses too much for stability in foods.

Because of the relatively high linolenate content in soybean oil, it is frequently blended with cottonseed, peanut, corn, or sunflower oil to lower the percentage of this acid. Soybean salad oil on the market today has a P/S ratio of 4.0.

Soybean oil is used extensively to make shortenings. When blended with cottonseed oil of 20 to 23% palmitic acid or the high levels in lard and tallow, a wide range of relatively inexpensive liquid or solid shortenings is produced.

Palm oil, a highly saturated oil, is one of the chief competitors of soybean oil. Large quantities of palm oil are being imported into the USA each year. This oil can be produced in greater quantities at less cost per hectare than any other edible oil. A hectare of palm trees produces nearly 4,036 kg (3,600 lb/A) of palm oil while soybeans produce about 336 kg (300 lb/A) of oil per hectare. The oil palm is also a perennial.

Lipid Utilization as Related to Fatty Acid Composition

The quality of fats and oils depends to a large extent on the use for which they are intended. Two general categories are the industrial or non-edible and the edible uses. The wide growth of the edible oil industry was due, in part, to the trend to more and more convenience foods. The edible uses of vegetable fats and oils far outweigh the non-edible uses.

High-quality vegetable oils for food are defined as having a high proportion of linoleic acid, an essential fatty acid, and a low proportion of linolenic acid. On the other hand, drying oils contain high proportions of the unsaturated fatty acid, linolenic, and produce films when exposed to the air. These oils are used in paints and other weatherproofing compounds.

If the oil has a high proportion of oleic or linoleic acid, no linolenic acid, and no more than 10 to 15% saturates, it will make a high-quality salad oil with no stability problem. Figure 5.3 presents a flow scheme for edible oils used to make salad, cooking, and shortening oils and margarine or other vegetable fats.

By blending with other oils or fats a whole range of general and special purpose products have been developed from oils with some linolenic acid. To improve the plastic range, such as in shortening, oils with linolenic acid can be mixed with other oils or fats of high palmitic acid content. Wide ranges in melting points and solid contents are obtained by blending low cost oils with lard (20 to 28% palmitic acid) and tallow (24 to 32% palmitic acid). These products are normally kept under refrigeration when open to retard development of rancid flavors and odors that result from oxidation of the double bonds in the unsaturated fatty acids.

GENETIC EFFECTS ON LIPIDS

Most oilseed improvement places major emphasis on high-yielding cultivars that are both high in oil and protein content. Although historically considered to be oil crops, soybean, flax (*Linum usitatissimum* L.), peanut, safflower, sesame (*Sesamum indicum* L.), and most other oilseed crops contribute significantly to the world supply of vegetable protein. Numerous studies have shown that as oil content increases, protein content of most of these species decreases.

Genetic variability for oil content has been confirmed in nearly every cultivated seed crop in which large numbers of cultivars of the species have been screened. In the next series of examples it will be shown that oil quantity usually increases at the expense of other seed components or yield.

Heritability for yield in the soybean is relatively low, about 52% (Johnson and Bernard, 1963). Independent heritability estimates for protein and oil are about 75%. Selection for either one of these characters alone could therefore result in a significant increase in that component, but simultaneous selection for increase in both is difficult because of their negative correlation. Little overall progress has been made in increasing soybean oil content from 1960 to 1970 (Brim, 1973). The objective was to simultaneously increase yield, and oil and protein contents. The inheritance of soybean oil and protein contents appears to be polygenic, that is, no single-gene effects have been identified (Johnson and Bernard, 1963).

Inheritance of fatty acid composition of soybean appears to be relatively complex. Screening germplasm collections has indicated that variability exists for the unsaturated fatty acids; however, levels of linolenic acid were still above those preferred by the edible oil industry. With complex inheritance and the absence of a breeding source very low in linolenic acid, it is unlikely that a great deal of effort will be expended trying to alter soybean oil quality.

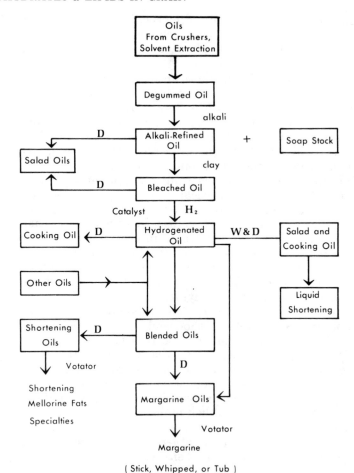

Fig. 5.3—Flow scheme for edible oil products, where D stands for deodorization and W for winterization. After Wolf and Cowan (1971).

A large portion of the genetic variability of safflower oil content is due to the difference in the hull-to-kernel ratio. Breeders have developed thin-hulled cultivars with hull percentage of 18 to 30%, about one-half the 40 to 50% range for commercial cultivars. As the proportion of hull-to-kernel decreases, oil contents of experimental strains have also increased. In a given cultivar of safflower, seed weight is highly positively correlated with oil content, and oil content is negatively correlated with the amount of hull.

In safflower cultivars, variations in iodine value (the quantity of iodine that combines at the double bonds of the unsaturated lipids) ranged from 87 to 149, mainly due to differences in the oleic and linoleic levels (Knowles, 1964). Oleic acid content varied from 9 to 87% and linoleic acid from 9 to 85%. Three allelic genes, *Ol*, *ol'*, and *ol* determine the proportions of linoleic and oleic acid (Knowles, 1968). Normal commercial cultivars have the genotype *Ol Ol* and their oil is used in paints and edible fats. Fine cooking and salad oil can be produced from the *ol ol* genotype which yields an oil similar to olive oil. The *ol' ol'* genotype has nearly equal linoleic and oleic acid composition.

Sesame is probably the most ancient oilseed used by man. There appears to be a great deal of variability of oil content among cultivars. Cultivars from the eastern Mediterranean produced the highest oil yields. Selection has been successful in increasing oil content within local cultivars throughout the area of sesame adaptation. High seed yield is correlated with late maturity and 2 to 7 genes appear to be involved in the inheritance of oil content (Culp, 1959).

The amount of oil in corn is primarily under genetic rather than environmental control. In a classical study of recurrent selection in corn, Illinois breeders shifted oil content from 4.69 to 17.98% in 70 generations (Dudley et al., 1974). The increase in oil content was attributed to increased germ size and a decrease in the amount of starch in the kernel. Continued increases in oil content demonstrate that many genes are involved for this factor. Breeding studies suggest that both relatively high-oil content (8%) and acceptable levels of linoleic acid (60%) can be combined in corn hybrids.

Oil of the castor bean (*Ricinus communis* L.) was recorded in Mesopotamian prescriptions thousands of years before the birth of Christ. Mass selection (selecting and bulking seed from similar plant types) was practiced on castor beans early in the history of agriculture. Uniform plants of a desired type can be grown by planting in areas isolated by a short distance from other castor beans.

Single-cross castor bean hybrids show considerable increase in yield of seed compared with their highest yielding parent. Heterosis or hybrid vigor is significant when one parent possesses dominant gene(s) for high-seed number and the other possesses dominant gene(s) for high-seed weight. However, seed size and weight are negatively correlated with oil content.

Breeding efforts are in progress to obtain peanut cultivars with a lower linoleic acid content to provide more flavor stability in the whole nuts. Spanish-type peanuts contain higher percentages of polyunsaturated fatty acids and higher total amounts of saturated fatty acids while Runner and Virginia-types of peanuts are higher in monounsaturated fatty acids, chiefly oleic acid. Peanut oil content appears to be an exception to the component substitution observed for many other crops. Plant breeders have increased seed-oil percentage and continue to increase total yield and oil yield per hectare.

Genetic variability has been found and the inheritance of oil quantity and quality determined for many oilseed and feed grain crops. Without a genetic breakthrough it appears unlikely that oil yield can be increased significantly while maintaining or increasing other seed components and yield. This phenomena has been described as an energy plateau. Peanut is among the few crops steadily increasing in energy harvested per hectare. Development of specialty crops, such as high-oil corn or other feed grains, on the other hand, would mean higher energy per unit of feed for the consuming animal.

ENVIRONMENTAL EFFECTS ON LIPIDS

Critical investigation of the effect of specific environmental variables on lipids was not generally possible until plant growth chambers were used to separate the interactions of the several environmental factors associated with field conditions. Improved analytical tools, such as gas liquid chromatography (GLC) and nuclear magnetic resonance (NMR), have allowed the scientist to examine the lipid quality of small samples. With NMR the oil content of a single seed can be accurately and rapidly measured without the destruction of that seed. Plant breeders are using NMR to increase oil yields in several species.

Field studies of soybean cultivars grown at a number of locations for several years were used to determine the association between oil percentage and minimum and maximum temperatures (Howell and Cartter, 1953). Temperature had its greatest effect on oil content during 20 to 30 and 30 to 40 days before maturity.

Nearly 90% of the soybean oil at maturity is accumulated within 40 days after flowering. This shows that oil production is equally competitive with other products the first 40 days after flowering. The relative ratios of oleic, linoleic, and the saturated fatty acids also reach an equilibrium in approximately 40 days; however, total oil and oil composition change greatly from flowering to maturity (Fig. 5.4 and 5.5). Thus the final composition of soybean oil must be achieved through differential rates of synthesis and/or incorporation of the fatty acids on the glycerol molecule.

Soybean oil quality also shows considerable variability when trials are conducted in different environments. Soybean oil analyzed from cultivars grown in 6 to 14 different environments ranged from 5 to 11% linolenic acid, 43 to 56% linoleic acid, 15 to 33% oleic acid, and the saturated acids from 11 to 26% (Collins and Sedgwick, 1959).

The variations in oil content and iodine value of some crops grown at several geographical locations are shown in Table 5.8. Differences in quality due to location appear to be small for crops such as corn or rape (*Brassica napus* L.). Flax and sunflower seed, on the other hand, showed considerable variation in iodine value. The iodine value for flax increased from 172 at location A to 202 at location E in the USA, nearly an 18% difference. This variation was not due to cultivar, as the same cultivar was grown at each site. One could also expect year-to-year variation in oil quality and quantity.

Under controlled conditions, there is an almost linear increase of oil in soybean seeds to increases in temperature (Howell and Cartter, 1958). Temperatures during the podfilling stage of 29, 25, and 21 C (84, 77 and 70 F) produced soybean seeds with 32, 21, and 19% oil, respectively. Temperature affected mainly the triglyceride fraction of the crude oil.

Table 5.8—Variations in oil content and iodine value of crops grown at several geographical locations.†

Location	Oil content‡					Iodine value‡				
	Flax	Sun-flower	Soy-bean	Corn	Rape-seed	Flax	Sun-flower	Soy-bean§	Corn§	Rape-seed
			%					mg#		
A	37	27	16	4.4	43	172	111	126	117	105
B	44	23	15	4.7	44	177	101	124	114	107
C	39	27	13	4.9	35	178	114	124	112	102
D	--	27	21	5.1	--	195	128	--	116	103
E	--	32	18	5.1	--	202	131	--	113	100

† Data compiled from the literature by C.D. Dybing, USDA, SEA-AR, Brookings, SD, unpublished.
‡ Iodine value and oil content not necessarily for the same sample or experiment.
§ Theoretical iodine values calculated from fatty acid composition data.
# (mg = 1.543×10^{-2} grains or 3.537×10^{-5} oz).

Table 5.9—Effect of temperature on the oil content of several oilseed crops.†

Species	Oil content (dry weight)			
	10 C (50 F)	16 C (60 F)	21 C (70 F)	26.5 C (80 F)
		%		
Rape	51.8	45.5	43.8	32.2
Sunflower	36.8	37.3	40.4	36.4
Safflower	25.4	26.0	25.4	23.8
Flax	46.6	43.1	40.3	35.1
Castor bean	--	48.0	51.2	46.5

† Data from Canvin (1965).

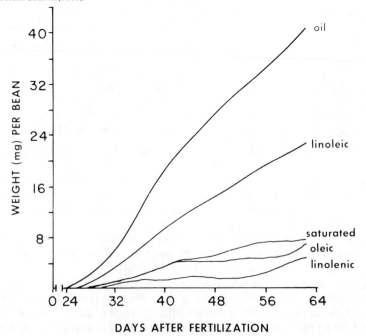

DAYS AFTER FERTILIZATION

Fig. 5.4—Changes in oil and lipid content (average weight in mg per bean) in field-grown Lincoln soybeans during maturation. Adapted from Simmons and Quackenbush (1954).

The effect of temperature on oil quantity from fertilization to maturity of rape, sunflower, safflower, flax, and castor bean is shown in Table 5.9. Oil contents of flax and rape were highest at the lower temperature of 10 C (50 F) and decreased with increasing temperature. Safflower was apparently stable within the range of temperatures used in the study, because the oil content did not vary significantly. Sunflower and castor bean seeds were curvilinear in their response to temperature, with the highest oil content at 21 C (70 F) and lower oil levels at higher and lower temperatures.

Oil quality, as indicated by fatty acid composition, from the castor bean and safflower was not affected by the change in temperatures in D. T. Canvin's study (1965). In sunflower and flax, as temperature increased, there was a decrease in the highly unsaturated fatty acids. The quantity of linoleic acid in sunflower and linolenic acid in flax seeds was almost proportionally offset by an increase in oleic acid as temperature increased. Erucic acid in rapeseed was at its highest level at 16 C (61 F) and oleic was at its lowest. These two fatty acids varied

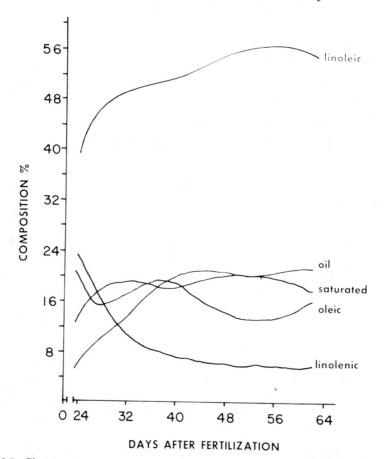

Fig. 5.5—Changes in percentage composition of soybean oil in developing beans. Adapted from Simmons and Quackenbush (1954).

Table 5.10—Percentage increase or decrease in iodine value and oil content of flax seeds exposed during seed development to varied levels of an individual environment component with other components held constant.†

| Component tested | Levels compared‡ | | Change§ | |
	Check	Maximum	Iodine value#	Oil content#
			%	
Air temperature (C)	15 (59 F)	30 (86 F)	− 21	− 18
Light μE m⁻² sec⁻¹ (400 to 700 nm)	175	375	+ 7	+ 4
Photoperiod (hr)	13	19	+ 5	+ 6
Atmospheric CO₂ (ppm)	325	1,270	0	0
Humidity (%)	60	70	0	0
Nitrogen (ppm)	112	224	− 4	− 10
Osmotic potential (bars)	− 0.5	− 10	0	− 6
Root-zone temperature (C)	15 (59 F)	25 (77 F)	+ 3	+ 7

† Data from C. D. Dybing, USDA, SEA-AR, Brookings, SD, unpublished.
‡ Standard environment: day temperature 25 C (77 F); night temperature 20 C (68 F); light 366 μE m⁻² sec⁻¹ (400 to 700 nm); photoperiod 16 hr; humidity and atmospheric composition uncontrolled.

§ % change = $\dfrac{x_{max} - x_{check}}{X\ check} \times 100.$

\# A value of 0 means no statistically significant (5% level) change was observed.

inversely and nearly proportionally as the temperature increased or decreased from 16 C (61 F).

Several environmental factors may affect flax seed oil quality and quantity (Table 5.10). The iodine value was 21% lower in oil from seeds grown at 30 C (86 F) than for those grown at 15 C (59 F). High light intensity, long photoperiod, and warm root temperature slightly increased the degree of lipid unsaturation. A high level of N slightly reduced the iodine value, although the plants yielded 40% more seed than the control level.

Oil quantity of flax seed was markedly decreased by warm air temperature, osmotic stress, and high N fertilization (Table 5.10). Long photoperiod, high light intensity, and warm root-zone temperature increased oil content. The factor affecting flax oil quantity and quality the most was air temperature.

The effect of temperature on oil content and iodine value of flax seeds appears to be greatest during the first 2 weeks of the flowering period (Dybing and Zimmerman, 1965). Results for corn generally agree with the observation on flax, in that warm temperature decreases both oil content and degree of unsaturation. The corn embryo, the main site of oil storage in the kernel, is more affected by temperature than is the endosperm.

Diseases may also affect oil quantity. Sunflower seeds, or achenes, harvested from groups of plants infected with four different pathogens were lower in oil content than achenes harvested from healthy plants (Zimmer and Zimmerman, 1972). Of the four diseases, all caused a decrease in oil content without influencing oil quality. Only a wilt pathogen affected oil content without increasing the hull percentage of the achenes.

In summary, environmental effects on quantity and quality of lipids generally occur in the developing seed only within the few weeks between flower fertilization and maximum dry weight accumulation.

Air temperature seems to affect lipids more than other factors. The reduction in oil content and in lipid unsaturation as ambient temperature increases appears to be an almost universal effect in seed plants. The soybean is a major exception. In soybean cultivars, oil content increases significantly as air temperature increases. High light intensity, long photoperiod, and warm root-zone temperature each increase flax oil content and lipid unsaturation. Plant diseases can also affect lipids.

Improved Vegetable Oils

An association of polyunsaturated fat content in the diet and reduced heart disease and blood cholesterol levels in humans has demanded that the quality of oil products be defined. Reducing saturated fatty acids and increasing polyunsaturates in the diet lowers the cholesterol content of the blood. The typical American diet has a (P/S) ratio of 0.3. A diet with a P/S ratio above 1.0 significantly lowers blood cholesterol.

The edible fat and oil industry has modified products to increase the amount of polyunsaturates and decrease the amount of saturated fats. One method is blending and another winterizing. Oil is winterized by gradually cooling over a period of several days to about 6 C (43 F). Large crystals of fat, sometimes called vegetable stearine, are then separated from the oil. Polyunsaturates are principally of vegetable origin and polysaturates of animal origin.

Acceptability of feed by animals and palatability of numerous convenience food items for humans depends on the absence of rancidity of the added fats and oils. The unsaturated double bonds of linolenic acid are easily oxidized, causing rancidity of feed and foods alike. Oils with linolenic acid are differentially hydrogenated (adding H to the double bonds by using Ni as a catalyst in the presence of heat and pressure) to improve their keeping quality.

To improve the biological value of vegetable oil and oil stability, plant breeders are attempting to develop oilseeds with less linolenic acid and more oleic and linoleic acids. Industry can further improve edible oils, such as soybean oil, by selectively hydrogenating the double bonds of linolenic acid.

SUMMARY

Man can change the genetic makeup of a plant, but aside from irrigation, he can do relatively little to modify the plant's weather-induced environment. The environment of a growing plant changes from day-to-day and hour-to-hour. Stress at one physiological stage of development may not change the quantity or quality of carbohydrate or lipid potential even though stress at another stage could be critical or irreversible.

The synthesis of starch from sugar is approximately twice as efficient bioenergetically as the transformation of sugar to oil. On the other hand, lipids contain more than twice the energy of carbohydrates. With specific goals, plant breeders can shift or divert photosynthate to high yields of either oil or protein. Mankind will see many changes in agricultural plants and the processed foods, feed, and other products made from them. The photosynthetic process may appear to be inefficient compared with the total energy received from the sun, but all life and nearly everything man produces or accomplishes is directly or indirectly dependent on this process.

LITERATURE CITED

Andres, C. 1976. Sweeteners outlook. Food Process. 37:46–48.

Brim, C. A. 1973. Quantitative genetics and breeding. p. 155–186. *In* B. E. Caldwell (ed.) Soybeans: Improvement, production, and uses. Am. Soc. Agron., Madison, WI.

Canvin, D. T. 1965. The effect of temperature on the oil content and fatty acid composition of the oils from several oil seed crops. Can. J. Bot. 43:63–69.

Collins, F. I., and V. E. Sedgwick. 1959. Fatty acid composition of several varieties of soybeans. J. Am. Oil Chem. Soc. 36:641–644.

Cowan, J. C., C. D. Evans, H. A. Moser, G. R. List, S. Koritala, K. J. Moulton, and H. J. Dutton. 1970. Flavor evaluation of copper-hydrogenated soybean oils. J. Am. Oil Chem. Soc. 47:470–474.

Creech, R. G. 1968. Carbohydrate synthesis in maize. Adv. Agron. 20:275–322.

Culp, T. W. 1959. Inheritance and association of oil and protein content and seed coat type in sesame, *Sesamum indicum* L. Genetics 44:897–909.

Dudley, J. W., R. J. Lambert, and D. E. Alexander. 1974. Seventy generations of selection for oil and protein concentration in the maize kernel. p. 181–212. *In* J. W. Dudley (ed.) Seventy generations of selection for oil and protein in maize. Crop Sci. Soc. Am., Madison, WI.

Dybing, C. D., and D. C. Zimmerman. 1965. Temperature effects on flax (*Linum usitatissimum* L.) growth, seed production, and oil quality in controlled environments. Crop Sci. 5:184–187.

Fergason, V. L., and M. S. Zuber. 1962. Influence of environment on amylose content of maize endosperm. Crop Sci. 2:209–211.

Geddes, R., C. T. Greenwood, and S. MacKenzie. 1965. Studies on the biosynthesis of starch granules. Part III. The properties of the components of starches from the growing potato tuber. Carbohyd. Res. 1:71–82.

Glasziou, K. T., T. A. Bull, M. D. Hatch, and P. C. Whiteman. 1965. Physiology of sugar-cane. VII. Effects of temperature, photoperiod duration, diurnal and seasonal temperature changes on growth and ripening. Aust. J. Biol. Sci. 18:53–66.

————, and K. R. Gayler. 1972. Storage of sugars in stalks of sugar cane. Bot. Rev. 38:471–490.

Hartt, C. E. 1967. Effect of moisture supply upon translocation and storage of ^{14}C in sugarcane. Plant Physiol. 43:338–346.

Hitchcock, C., and B. W. Nichols. 1971. Plant lipid biochemistry. Academic Press, NY.

Hoover, M. W., and S. J. Harmon. 1967. Carbohydrate changes in sweet potato flakes made by the enzyme activation technique. Food Technol. 21: 115-118.

Howell, R. W., and J. L. Cartter. 1953. Physiological factors affecting composition of soybeans. I. Correlation of temperatures during certain portions of the pod filling stage with oil percentage in mature beans. Agron. J. 45:526-528.

————, and ————. 1958. Physiological factors affecting composition of soybeans: II. Response of oil and other constituents of soybeans to temperature under controlled conditions. Agron. J. 50:664-667.

Johnson, H. W., and R. L. Bernard. 1963. Soybean genetics and breeding. p. 1-73. *In* A. G. Norman (ed.) The soybean. Academic Press, NY.

Knowles, P. F. 1964. Variability in oleic and linoleic acid contents of safflower oil. Econ. Bot. 19:53-62.

————. 1968. Association of high levels of oleic acid in the seed oil of safflower (*Carthamus tinctorius*) with other plant and seed characteristics. Econ. Bot. 22:195-200.

Krog, N. E., D. L. Tourneau, and H. Hart. 1961. The sugar content of wheat leaves infected with stem rust. Phytopathology 51:75-77.

Laughnan, J. R. 1953. The effect of the sh_2 factor on carbohydrate reserves in the mature endosperm of maize. Genetics 38:485-499.

McOsker, D. E., F. H. Mattson, H. B. Sweringen, and A. M. Kligman. 1962. The influence of partially hydrogenated dietary fats on serum cholesterol levels. J. Am. Med. Assoc. 180:380-385.

Rackis, J. J., D. H. Honig, D. J. Sessa, and F. R. Steggerda. 1970. Flavor and flatulence factors in soybean protein products. J. Agric. Food Chem. 18: 977-982.

Rao, P. S. 1971. Studies on the nature of carbohydrate moiety in high yielding varieties of rice. J. Nutr. 101:879-884.

Shaw, H. R. 1953. An international glance at sucrose content of cane. Proc. Int. Soc. Sugarcane Technol. 8:283-289.

Simmons, R. O., and F. W. Quackenbush. 1954. Comparative rates of formation of fatty acids in the soybean seed during its development. J. Am. Oil Chem. Soc. 31:601-603.

Tsai, C. Y., F. Salamini, and O. E. Nelson. 1970. Enzymes of carbohydrate metabolism in the developing endosperm of maize. Plant Physiol. 46:299-306.

Waldron, J. C., K. T. Glasziou, and T. A. Bull. 1967. The physiology of sugarcane. IX. Factors affecting photosynthesis and sugar storage. Aust. J. Biol. Sci. 20:1043-1052.

Wolf, W. J., and J. C. Cowan. 1971. Soybeans as a food source. CRC Press, Cleveland, OH.

Woodwell, G. M. 1970. The energy cycle of the biosphere. Sci. Am. 223:64-74.

Zimmer, D. E., and D. C. Zimmerman. 1972. Influence of some diseases on achene and oil quality of sunflower. Crop Sci. 12:859-861.

SUGGESTED READING

Caldwell, B. E. (ed.). 1973. Soybeans: Improvement, production, and uses. Am. Soc. Agron., Madison, WI.

Inglett, G. E. (ed.). 1970. Corn: Culture, processing, products, AVI Publishing Co., Westport, CT.

Milborrow, B. V. (ed.). 1973. Biosynthesis and its control in plants. Academic Press, NY.

Sprague, G. F. (ed.). 1977. Corn and corn improvement. Am. Soc. Agron., Madison, WI.

Walden, D. B. (ed.). 1978. Maize breeding and genetics. John Wiley and Sons, NY.

Wall, J. S., and W. M. Ross (eds.). 1970. Sorghum production and utilization. AVI Publishing Co., Westport, CT.

Weiss, E. A. 1971. Castor, sesame and safflower. Barnes and Noble, Inc., NY.

Chapter 6

Genetic and Environmental Effects on Forage Quality

C. S. HOVELAND

Auburn University
Auburn, Alabama

W. G. MONSON

SEA-USDA
Tifton, Georgia

Forages supply digestible energy, protein, and minerals for maintenance, growth, and wool, beef, or milk production of ruminant animals. Performance of livestock depends on genetic potential of the animal, as well as availability and quality of the forage.

GENETIC EFFECTS

Forage species differ widely in their chemical composition and nutritive value, even when grown under the same environmental conditions. Also, individual plants within a species may differ genetically in nutritive value. Selection and breeding of more nutritious forages for improved animal performance are therefore possible.

Energy

Digestible energy generally determines intake of a forage species and ruminant performance. Most forages are inferior to concentrates as energy food sources. A given quantity of total digestible nutrients (TDN) of forage is usually less efficiently utilized than the digestible equivalent of concentrate. Hence the emphasis should be on improvement and maintenance of forages high in digestible energy (DE).

Temperate Grasses

In general, temperate grasses are more digestible than tropical species. This effect is under genetic control, but also may partially be a result of high temperatures and lower nutrient inputs in tropical grassland systems.

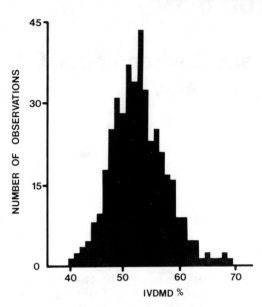

Fig. 6.1—In vitro dry matter digestibility (IVDMD) for 5-week-old forage from the world collection of bermudagrasses grown at Tifton, GA; 2-year average. (Burton and Monson, 1972).

Within the temperate grasses, genera and species may differ in digestible energy content (Minson et al., 1964). For instance, perennial ryegrass (*Lolium perenne* L.), annual ryegrass (*Lolium multiflorum* Lam.), oats (*Avena sativa* L.), wheat (*Triticum aestivum* L.), and rye (*Secale cereale* L.) are usually more digestible than orchardgrass (*Dactylis glomerata* L.), tall fescue (*Festuca arundinacea* Schreb.), or smooth bromegrass (*Bromus inermis* Leyss.) when compared at the same stage of maturity.

Even at the same stage of maturity temperate grass cultivars can differ in digestibility (Cooper et al., 1963; Sleper et al., 1973). Late maturing cultivars of ryegrass have been shown to maintain higher levels of digestibility to later dates than early-maturing cultivars (Raymond, 1969). Individual genotypes, within the same cultivar, can be more digestible than others. In orchardgrass, the digestible genotypes within a population maintained this advantage through successive growth stages in all plant parts (Julen and Lagar, 1966). Since these factors are genetically controlled, it is possible to select digestible types in a breeding program.

Sometimes no difference of consequence in nutritive value has been found among genotypes of certain species. Other criteria, such as yield and disease resistance, then are of primary importance to the grass breeder in improving yield of energy.

Small grain forage is succulent, palatable, and high in digestibility if utilized prior to full bloom stage. Research work at Tifton, Georgia showed average in vitro dry matter digestibility (IVDMD) values of

76, 74, 77, and 77% for oat, wheat, rye, and barley, respectively, over a range of fertility levels (Morey et al., 1969). Variation in digestibility among oat cultivars is not sufficient to warrant primary emphasis in a breeding program (Stuthman and Marten, 1972). Selection for yield potential is a better approach to improve energy production. Many other annual forage crops, such as legumes, which are normally of high quality but may possess other genetic factors which reduce utilization, may be exceptions.

Tropical Perennial Grasses

Dallisgrass (*Paspalum dilatatum* Poir.), kikuyu (*Pennisetum clandestinum* Hachst.), and johnsongrass [*Sorghum halapense* (L.) Pers.] are examples of warm season perennial grasses with higher levels of digestibility and animal performance than many others, such as bahiagrass (*Paspalum notatum* Flugge) and carpetgrass (*Axonopus affinis* Chase).

Of all forage plants, the tropical perennial grasses probably have the greatest potential for improvement of DE content. Plant breeders improved DE content of several species, such as kleingrass (*Panicum coloratum* L.), digitgrass (*Digitaria* sp.), and bahiagrass. The most impressive example is bermudagrass [*Cynodon dactylon* (L.) Pers.] (Burton and Monson, 1972). Several plants in the world collection of bermudagrass have IVDMD values between 65 and 70% (Fig. 6.1).

Thus, it should be possible to advance digestibility beyond that of currently used cultivars which have IVDMD values of 45 to 55%. While these more digestible plants do not have the combination of characters to make them directly usable as forage crops, they are useful as parents in a hybridization program. Problems are encountered in such research—the introduction with the highest IVDMD may lack winterhardiness, may have high concentration of prussic acid glycosides, and may rarely flower. Since the hybrid is rarely as good as the best parent in IVDMD, it is essential to identify highly digestible parents.

Hybrids between Coastal bermudagrass and a highly digestible cold-susceptible bermuda from Kenya were screened at Tifton for vigor, disease resistance, drought tolerance, yield, winterhardiness, and digestibility (Burton et al., 1967). One hybrid was consistently 12% more digestible than Coastal bermudagrass. Steers consuming the forage gained 30% faster than those eating Coastal bermudagrass. The resulting hybrid, 'Coastcross-1,' released in May 1967, was the first forage cultivar bred for improved quality using modern methods for digestibility screening (Lowrey et al., 1968). Digestibility of the new cultivar and animal production from it have consistently been superior to Coastal bermudagrass (Fig. 6.2). However, Coastcross-1 lacks cold-hardiness for much of the southeastern U.S.

Table 6.1—Performance of tall and dwarf pearl millet when grazed at Tifton, GA; 1968.†

Management	Tall	Dwarf
Steer days/ha	654	556
Steer gain, kg/ha (lb/a)	386 (344)	393 (350)
Average daily steer gain, kg (lb)	0.59 (1.30)	0.71(1.56)

† From Johnson et al. (1968).

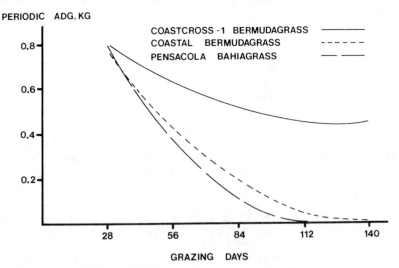

Fig. 6.2—Average daily gain (ADG) of yearling steers grazing three warm season perennial grasses for 140-day grazing season at Tifton, GA (Utley et al., 1974).

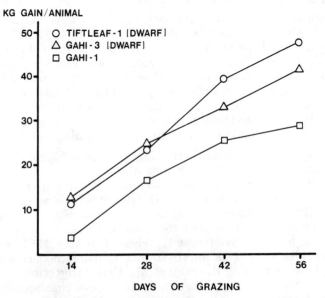

Fig. 6.3—Cumulative gains per animal from Tiftleaf-1, Gahi-3, and Gahi-1 hybrid pearl millets (Johnson et al., 1976).

Tropical Annual Grasses

A number of corn (*Zea mays* L.) single crosses and cultivars used for silage have been examined for genetic variation in forage quality traits (Roth et al., 1970). Genetic variability was found for IVDMD, acid detergent fiber, acid detergent lignin, cell wall constituents, and crude protein (CP). Breeding corn cultivars for silage quality superior to those developed for grain yield alone is feasible.

In corn, a brown midrib mutant is associated with reduced lignin content and improved IVDMD (Muller et al., 1971). Brown midrib and other genetically controlled characters potentially could be used to improve energy from forage.

Sorghum [*Sorghum bicolor* (L.) Moench], sudangrass [*Sorghum sudanense* (Piper) Stapf.], pearl millet [*Pennisetum americanum* (L.) Leeke], and sorghum × sudangrass hybrids are widely used as summer annual forage crops. All are capable of producing good yields of high quality forage when grown in their area of adaptation. Digestibility varies among genotypes, and may be improved. Selecting for increased IVDMD in pearl millet is possible because this trait is heritable (G. W. Burton and W. G. Monson, unpublished data).

A high leaf:stem ratio is associated with good forage quality of most species. This ratio can be drastically changed in pearl millet by the introduction of the dwarf gene, which shortens the stem internodes without otherwise altering the plant. The net result is nearly a 50% reduction in stem tissue. The effects of the dwarf gene on forage production and quality has been demonstrated both in the laboratory and in animal feeding studies in Georgia. In spaced plants, the dwarf gene reduced internode length, plant height, and dry matter yields and increased leaf percentage, IVDMD, and CP content of stems (Burton et al., 1969). When fed chopped, dehydrated, dwarf pearl millet, dairy heifers ate 21% more, gained 49% faster, and produced as much gain per hectare as that from similarly-treated tall pearl millet. Grazing trials showed similar results (Table 6.1). Thus, it appears possible through the introduction of the dwarf gene to significantly increase the concentration of energy and protein in an annual forage plant. Dwarf hybrids should produce significantly more forage than the inbred lines tested. 'Tiftleaf-1,' a dwarf pearl millet, has improved animal performance when grazed (Fig. 6.3). If quality can be increased in this manner, it would be easy to improve the quality of other annual grass species by introducing a dwarf gene into a cultivar through backcrossing.

Breeding to improve nutritive value should not emphasize solely leafiness and fineness of stem. For example, increased stem diameter and more pith parenchymea, which is largely non-lignified cellulose, may increase IVDMD of sorghum (Burns et al., 1970).

Legumes

Legumes generally are high quality forages. Perennial legumes, such as alfalfa (*Medicago sativa* L.), red clover (*Trifolium pratense* L.), birdsfoot trefoil (*Lotus corniculatus* L.), and white clover (*Trifolium repens* L.) are highly digestible, and can be compared to annual clovers, such as crimson (*Trifolium incarnatum* L.), arrowleaf (*Trifolium vesticulosum* Savi), and subterranean (*Trifolium subterraneum* L.). Daily gain of beef cows and calves is improved when clover is present in pastures (Table 6.2). Steer gains have been similar from clover and well-fertilized cool season, annual grass pastures.

Digestibility of legumes is generally good, with the added advantage that animals eat more of them than grasses. What, then, are the genetic factors affecting energy production and utilization? First legumes vary genetically in dry matter yield and pest resistance. Both are primary factors in the production of energy. Yield, a traditional selection factor in breeding, will be mentioned briefly here. Alfalfa yields up to 22 metric tons/ha (9.8 T/a). While a good portion of this increase is due to improved management practices, which will be discussed later in this chapter, such yields would not be possible without genetically improved cultivars (Gil et al., 1967).

Crude fiber content of individual alfalfa plants differs considerably (Fig. 6.4). Thus digestibility of alfalfa may be increased by selection. However, because alfalfa is already highly digestible, selecting for increased forage yield could pay larger dividends than selecting for digestibility.

Tropical legumes such as kudzu (*Pueraria lobata* (Willd) ohw), stylo (*Stylosanthes humilis* H.B.K.), and siratro (*Macroptilium atropurpureus* O.C.) also usually enable a much higher animal intake of DE than the grasses. Australian researchers have found that several tropical legumes maintain good digestibility and are eaten even after frost damage and death of the plants (Milford, 1967).

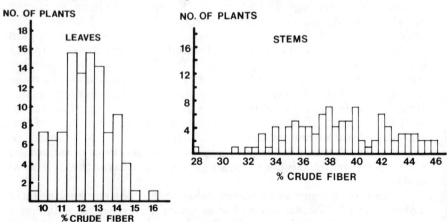

Fig. 6.4—Frequency distribution of 100 alfalfa plants for crude fiber content (Heinrichs and Troelson, 1965).

Table 6.2—Examples of beef animal performance on pastures of grasses vs grass-legume mixtures.

Location and reference	Years of trial	King of pasture	Kind of animal	Daily gain per head	
				kg	lb
Alabama Hoveland et al. (1978)	3	Bermudagrass	Cows	0.22	0.48
		Bermudagrass-arrowleaf & crimson clovers		0.62	1.36
		Bermudagrass	Calves	0.71	1.56
		Bermudagrass-arrowleaf & crimson clovers		0.89	1.96
Alabama Harris et al. (1972)	8	Tall fescue	Steers	0.60	1.32
		Tall fescue-ladino clover		0.66	1.46
		Orchardgrass		0.80	1.76
		Orchardgrass-ladino clover		0.83	1.83
Indiana Smith et al. (1975)	3	Tall fescue	Cows	0.05	0.11
		Tall fescue-ladino & red clovers		0.34	0.75
		Tall fescue	Calves	0.59	1.30
		Tall fescue-ladino & red clovers		0.82	1.81
Virginia Blaser et al. (1969)	10	Orchardgrass	Steers	0.48	1.06
		Orchardgrass-ladino clover		0.58	1.23

Barriers to Availability of Energy in Forages

Various physical features of forages are barriers to digestion by rumen microorganisms (Monson et al., 1972). The in vitro digestion of green bermudagrass and pearl millet leaves for 12, 48, and 96 hours in rumen fluid indicate that the anatomical features of the leaf affect rate of digestion. The relative amounts of the leaf made up by vascular bundles, cutinized and lignified cell walls, mesophyll, and other anatomical structures vary among cultivars within a species (Hanna et al., 1973). Cell types could be tested in screening plant materials for those readily digested. Rumen microorganisms consistently digest the larger and less compactly arranged cells of the mesophyll first, while bundle sheath cells and cutinized or lignified cells remain undigested after 96 hours (Fig. 6.5).

The scanning electron microscope elucidates the microanatomical features affecting forage digestion (Fig. 6.6). Digestibility of forages may be improved through selection of plant types with favorable microanatomical features.

Protein

Protein content frequently is used as a measure of forage quality. While the percent CP (%N × 6.25) in a given forage may not always reflect overall quality, any good forage must contain adequate protein. The amount of protein in forages is subject to environment and management.

Digestibility of protein relates positively to the concentration of protein in the plant. Hence, the digestibility of the protein is not generally determined in breeding programs. The percentage of digestible protein in the forage fed may be computed from the equation, $Y = 0.929 \times -3.48$, where Y = digestible crude protein (% of forage DM) and x = CP % in forage (Holter and Reid, 1959). The amount of protein is important in overall forage quality, however, and levels lower than 7 or 8% may have adverse effects on IVDMD and/or intake of a forage (Milford and Minson, 1963).

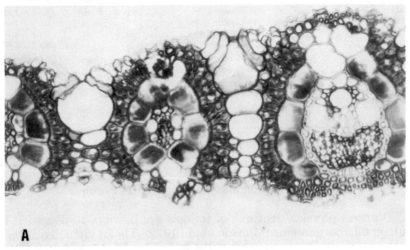

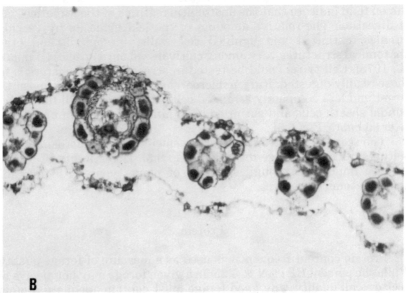

Fig. 6.5—Cross section of Coastal bermudagrass leaf after 0 (A) and 96 hours (B) digestion in rumen fluid (Hanna et al., 1973).

Grasses

The temperate grasses often contain more protein than tropical grasses. However, this difference may depend more on management and environment than on inherent differences among species.

Protein content among tropical perennial grass genotypes may inherently differ. For example, Coastcross-1, specifically bred for higher dry matter digestibility also is usually about 2 percentage units higher in CP content than is Coastal bermudagrass.

Legumes

Legumes usually contain more protein than do grasses that are not heavily fertilized with N. Thus, a legume growing with a low-protein grass can supply the requirements of growing animals and also supply some N to the grass.

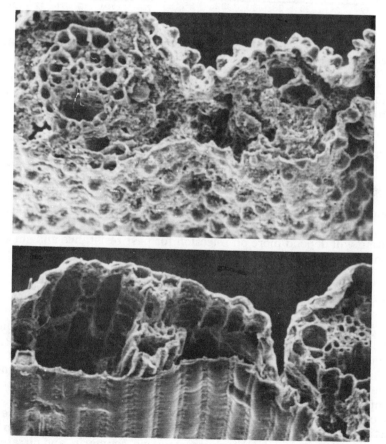

Fig. 6.6—Photomicrographs of leaf blade sections digested in rumen fluid for 12 hours. Tissues of Kentucky 31 tall fescue (lower) are more digested than Coastal bermuda (upper) × 750. (Akin et al., 1973).

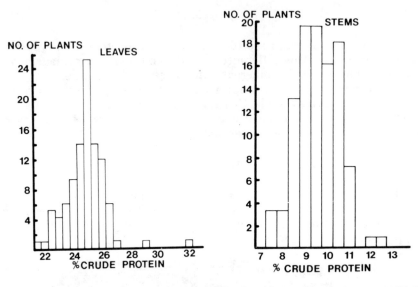

Fig. 6.7—Frequency distribution of 100 alfalfa plants for crude protein content (Heinrichs and Troelson, 1965).

Clonal selection for higher protein concentration in all plant parts of alfalfa is possible (Gengenbach and Miller, 1972). However, analysis of total herbage would best accomplish selection for protein. Variability for CP exists in a segregating alfalfa population (Fig. 6.7), and the variation is sufficient enough so that selection for increased CP could be made.

No single tropical legume is as widely adapted and used as alfalfa in the temperate zone, even though numerous tropical legume species are available. In many areas, dependable summer-growing perennial legumes are difficult to find. Disease, insect problems, and lack of winterhardiness are frequent problems. Thus, production of DE and protein will depend more on selection for pest resistance and survival than on breeding for digestibility and protein content.

Other Factors Affecting Forage Quality

Compounds occurring in forage crop plants can affect palatability, intake potential, or digestibility of the forage by livestock. They are usually secondary metabolites and may affect animal response directly or indirectly (Barnes and Gustine, 1973). Many of them are under genetic control. Adequate analytical methods are essential for these compounds. Diverse plant populations need to be developed so that selection for desired amounts of these compounds can be made. A partial list of these antiquality factors would include alkaloids, tannin, saponin, prussic acid (HCN), phytoestrogens, coumarin, and nitrate.

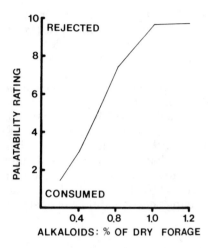

Fig. 6.8—Relationship of palatability rating (high rating = low palatability) and alkaloid concentration in reed canarygrass (Simon and Marten, 1971).

Not all anti-quality compounds directly alter energy and protein intake, or utilization; only alkaloids and tannins have demonstrated effects. The others affect animal health or reproduction and, indirectly, performance. In severe cases, death may result.

Mineral uptake is at least partially under genetic control, and an imbalance in the forage may affect animal performance. Mineral content of forage is also strongly influenced by fertilization and other management practices. Mineral imbalances may be expressed in the ruminant animal by a reduced growth rate or by an acute syndrome, such as tetany.

Alkaloids

Alkaloids are present in high amounts in relatively few of the vascular plants, although a rather large number of alkaloids have been identified. Forage plants containing alkaloids in significant amounts are represented by reed canarygrass (*Phalaris arundinacea* L.), tall fescue, and lupine (*Lupinus angustifolius* L.).

Eight indole alkaloids have been identified in reed canarygrass, with gramine and tryptamines predominant (Marten, 1973). The effects of the alkaloids appear to be more related to palatability and intake than to digestibility (Fig. 6.8). The result is lower energy intake and lowered animal performance. Alkaloids in reed canarygrass do not seem to affect digestibility. In contrast, alkaloids in phalaris (*Phalaris aquatica* L.) in Australia may cause "staggers," lack of muscle control, and eventual death of sheep.

The heritability of total alkaloid concentration in reed canarygrass is quite high. Research work in Minnesota (Marten, 1973) indicates that rapid gains can be made in early cycles of selection for low alkaloid concentration. Thus, it is feasible to breed reed canarygrass cultivars with lower levels of alkaloids and improve animal performance.

Alkaloids have also been indicated as the cause of the low performance syndrome often observed during late summer with cattle grazing tall fescue (Bush and Buckner, 1973). Daily gains of growing steers are commonly only 0.3 to 0.5 kg (0.7 to 1.1 lb) on tall fescue pasture as compared to 0.6 kg (1.3 lb) on orchardgrass. This performance often parallels increasing alkaloid concentrations. Unlike reed canarygrass, an alkaloid in tall fescue, perloline, inhibits rumen cellulose digestion (Bush et al., 1970). It is thought that alkaloids inhibit rate of digestion in the rumen, decreasing the availability of energy and nutrients to the animal.

Tannins

Tannins are usually associated with a bitter taste in plant tissues. They are present in a number of forages and have been associated with lowered intake and digestibility (Barnes and Gustine, 1973; Donnelly and Anthony, 1969). A notable breeding success has been sericea lespedeza [*Lespedeza cuneata* (Dumont G. Don)]. Selection for low tannin in this species has been successful. Despite some reported beneficial effects of tannins for bloat control and fungicidal activity, selection for low tannin content has led to improved quality sericea (Donnelly and Anthony, 1970).

Several other legumes also contain tannins in significant amounts. Birdsfoot trefoil is not generally reduced in quality because of tannins, although cattle may temporarily avoid it if given a choice. Modest levels of tannin apparently prevent or reduce bloat in cattle grazing birdsfoot trefoil, crownvetch (*Coronillus varia* L.), arrowleaf clover, and other non-bloating legumes (Hoveland et al., 1972; Jones and Lyttleton, 1971). A beneficial effect of tannins may be as a protein binding agent to protect protein from microbial degradation in the rumen and thus to increase N utilization.

Saponins

Saponins are glycosides that occur in some forage plants, such as alfalfa. They alter the surface tension of the rumen contents, trapping bubbles of fermentation gases, and contributing to bloat in ruminant animals. Alfalfa cultivars differ in content and kind of saponins (Hanson et al., 1973).

Cyanogenic Glycosides

Cyanogenic glycosides occur in some forages, such as sorghum. Large amounts of prussic acid (HCN) may be released in a short time when plant tissue is subjected to stress, such as drought or freezing (Barnes and Gustine, 1973). The HCN is readily absorbed into the blood of grazing ruminants, causing asphyxiation at the cellular level, and often death of the animal. Sorghum genotypes may differ in cyanogenic potential.

Coumarin

Sweetclover (*Melilotus* sp.) contains glycosides which, under wilting, maceration, or freezing, are converted to dicoumarol by microbes during spoilage (Barnes and Gustine, 1973). The dicoumarol prevents blood from clotting, resulting in bleeding and possible death of animals ingesting the forage. Low-glycoside sweetclover genotypes have been isolated and low courmarin cultivars are now available (Cooper, 1973).

Estrogens

Estrogenic flavonoid compounds, occurring in legumes such as alfalfa, and white, red, and subterranean clovers, may interfere with reproduction in sheep and cattle (Barnes and Gustine, 1973). Wide differences in estrogen level have been found among lines of subterranean clover (Cooper, 1973). In fact, subterranean clover cultivars with low estrogenicity are currently available in Australia.

Minerals

Generally, mineral content of forage is adequate for ruminant nutrition when fertilizer is applied to obtain satisfactory yields. Legumes are higher in Ca than grasses. There are some exceptions, however. On some highly leached tropical sandy soils, grasses may be low in Ca and P. Conversely, on some organic soils, legumes may accumulate excessive and toxic levels of Mo. In the U.S. Northern Great Plains, certain forage plants may accumulate toxic levels of Se.

Grass tetany is a metabolic disorder of lactating cows when intake of Mg is too low (Grunes, 1973). The problem is generally confined to grass pastures, because legumes normally have adequate or high levels of Mg. Fertilization will not insure sufficiently high levels of Mg or prevent grass tetany. Plant breeders have found a wide range in Mg uptake among grass genotypes, a finding that should result in development of tetany-resistant grass cultivars.

Special Animal Problems Related to Forage Quality

As livestock production under grazing is intensified, several detrimental forage-animal stresses may develop (Reid and Jung, 1973). The increased use of the high quality legumes in grazing may increase the incidence of bloat. Highly fertilized winter annual grasses, which are often grazed during cool wet periods, may lead to increased occurrence of grass tetany. Heavy N fertilization of tall fescue may increase fescue toxicosis and fescue foot in cattle. These types of production disorders present special challenges to: 1) the forage-livestock manager who must try to minimize or treat these maladies and 2) the forage breeder who must seek improved cultivars less apt to cause these responses in the grazing animal.

ENVIRONMENTAL EFFECTS

The genetic potential of forage species to produce high quality forage may be greatly modified by environment. In a broad sense, the environment not only includes climate, over which the forage producer generally has little control, but also grazing management, time of cutting, fertilization, burning, and pest control, all of which offer many opportunities to influence forage quality.

Climate

Forage species are generally chosen for their climatic adaptation and seasonal dry matter production, rather than for their nutrient content to supply the needs of ruminant animals. As noted previously, forage species differ greatly in their chemical composition. In both temperate and tropical environments, legumes generally contain more Ca and Mg than do grasses, and digestibility and intake are also higher. Legumes contain more protein than do temperate grasses fertilized with less than 200 kg/ha N annually. Temperate or cool season perennial grasses are generally more digestible and contain more protein and minerals than do tropical species, but this may be partially the result of climate and management.

Temperature

Higher temperatures increase the metabolic activity of forage plants, resulting in a rapid turnover of energy and synthesis of new cells during rapid growth. Thus, there is an accumulation of cell wall structures. These contain lignin, which is not digestible. Lignification also reduces the availability of total non-structural carbohydrates (TNC) of timothy (*Phleum pratense* L.) at high temperature (Table 6.3).

Table 6.3—Total non-structural carbohydrates (percent of dry weight) in timothy plants at early anthesis following growth at cool (18/10 C) and warm (32/23 C) day/ night temperatures with reversal of temperature regimes at inflorescence emergence.†

Plant part	Temperature regimes			
	Cool	Cool-warm	Warm	Warm-cool
		%		
Inflorescences	9.5	10.5	10.7	11.0
Leaf blades	16.4	7.0	8.1	10.1
Stems and sheaths	16.6	7.2	8.3	15.6
Stubble (7.6 cm)	25.5	17.6	19.0	19.0
Roots	8.4	3.2	6.5	8.2

† From Smith (1970).

Table 6.4—Composition of alfalfa forage harvested at first flower following growth under cool (18 C day/10 C night) and warm (32 C/24 C) temperatures.†

Constituent	Cool	Warm
	%	
IVDMD‡	66.5	64.0
TNC§	12.7	5.4
Crude protein	17.0	21.3
P	0.24	0.34
K	1.34	2.35
Ca	1.89	1.55
Mg	0.49	0.35

† From Smith (1969).
‡ In vitro dry matter digestibility.
§ Total non-structural carbohydrates.

However, if high temperatures, up to inflorescence emergence, are followed by cooler temperatures until harvest at early anthesis, then TNC increase. In this case, hay quality should be higher.

Both temperate and tropical grass digestibility have been shown to decline with increasing temperatures (Deinum et al., 1968; Minson and McLeod, 1970). In climates where growth continues during mild winters, forage will contain lower amounts of cellulose than later in the season when temperatures are higher. Non-structural carbohydrates or sugars and protein are normally high in winter-growing cereal grasses and annual ryegrass.

Various nutrients of alfalfa are affected by high temperature (Table 6.4). High temperatures reduce the forage concentration of IVDMD, TNC, Ca, and Mg, while CP, P, and K are increased. The crude protein increases at high temperature because of higher leaf percentage as well as increased CP in the leaves (Vough and Marten, 1971).

In forages such as sericea lespedeza, high temperatures accelerate the accumulation of tannins. High amounts of tannin decrease forage digestibility. Thus, animal performance on sericea lespedeza is better in spring than in summer, even when forage is at the same stage of maturity.

C. S. HOVELAND & W. G. MONSON

Table 6.5—Non-structural carbohydrate concentrations (percent of dry weight) in perennial ryegrass forage after 4 weeks of growth in three different temperature and light regimes.†

	Light		
Day/night temperatures	490	350	90
		cal/cm²/day	
25/20	21.2	18.8	8.2
20/15	26.7	21.2	7.9
15/10	33.2	28.4	9.0

† From Deinum (1966).

Temperature has little effect on alkaloid content of reed canary-grass (Marten, 1973). For a few days after freezing temperatures, large quantities of toxic HCN from cyanogenic glycosides are present in sorghum and sudangrass forage.

Light

In the field, it is often difficult to separate the effects of light and temperature on forage nutritive quality. Since light increases photo-synthesis, improvement in digestibility of the forage can be expected as sugar content increases. The content of TNC in forage may be de-creased under low light (Table 6.5). Light values below 350 cal/cm²/day are common during cloudy spring weather in temperate high latitude areas. Similar reductions in TNC of bermudagrass have been found in the southeastern U.S. Shading of one species by another species within the crop canopy may also have an adverse effect on TNC levels.

Soil Moisture

The major effect of water stress on forage production is a reduc-tion in yield. A secondary effect is reduction in water content of the forage. Under arid conditions, pasture can supply only a small portion of the animal's water requirement. With lush green pasture containing 80 to 85% water, the ruminant will consume 4.5 to 6 kg of water for every kilogram of dry forage eaten.

The effects of water stress on nutrient content is less clear. Protein content may increase as growth and assimilation of carbohydrates are suppressed more than N uptake by the plant. Nitrogen may accumu-late as nitrates, causing potential toxicity problems in some grasses. Water stress generally hastens maturity, having little effect on DE of the forage. Digestibility of alfalfa may be enhanced by drought, partly because of increased leaf-stem ratios due to stunting (Vough and Marten, 1971). In other species, particularly in more mature tissues, maturity is accelerated with premature senescence and decreased di-gestibility of the forage. Drought may result in loss of leaves in legumes, reducing digestibility of the forage. Mineral content of

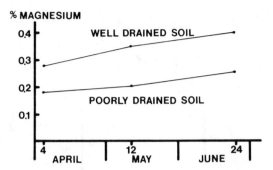

Fig. 6.9—Magnesium in tall fescue forage as affected by soil drainage in the Alabama Piedmont (Elkins et al., 1978).

forages is generally unaffected by water stress except for a decrease in P content.

Water stress may adversely affect forage plants in other ways. Alkaloids are increased in reed canarygrass under drought (Marten, 1973). Cyanogenic glycosides in sorghum often release toxic levels of HCN during drought.

Excess soil water in poorly drained soils during late winter and spring results in very low levels of soil oxygen. This situation may reduce the uptake of Mg by tall fescue and ryegrass to below 0.2 to 0.25% (Elkins et al., 1978), low enough to cause tetany in cattle grazing the forage (Fig. 6.9).

Stage of Maturity

Most forage species decline in nutritive value with age. This is a result of an increasing ratio of stem to leaf tissue combined with increasing lignification of cell walls. Stage of maturity has a greater influence on nutritive quality of forage than any other factor. Fortunately, this factor can be controlled by skillful grazing management and timing of hay cuts.

Although dry matter yield of alfalfa increases to the full bloom stage, digestibility and CP decline after the early bud stage (Fig. 6.10). High quality alfalfa hay is especially important for lactating dairy cows. Declining digestibility from advancing maturity can be expected to reduce milk production, because of reduced DE and lower voluntary intake of the forage.

The effect of cutting date on voluntary intake of alfalfa and orchardgrass by lambs in West Virginia illustrates the importance of this management variable (Fig. 6.11). Even at the seed stage in July or August, the relatively high intake of alfalfa provided sufficient productive energy for reasonably good liveweight gain. By contrast, intake of first growth orchardgrass declined rapidly with maturity; by late bloom, the forage supplied little more than maintenance energy to the animals.

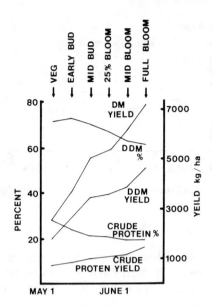

Fig. 6.10—Changes in dry matter (DM) yield, percentage, and yield of digestible dry matter (DDM), and percentage and yield of crude protein at different growth stages of Vernal alfalfa (Barnes and Gordon, 1972).

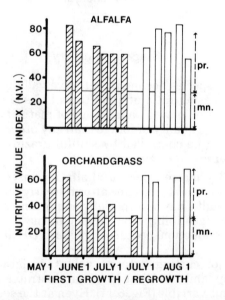

Fig. 6.11—Influence of cutting date on the intake of digestible dry matter (Nutritive Value Index) from alfalfa and orchardgrass. The Nutritive Value Index expresses the combined effect of energy digestibility and relative intake. The area below the horizontal line is the ingested energy calculated to maintain body weight. The area above the line represents the amount available for production.
(Barnes and Gordon, 1972).

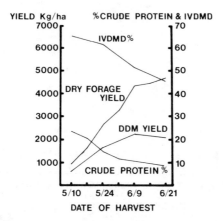

Fig. 6.12—Yield, in-vitro dry matter digestibility (IVDMD), and crude protein content of Lincoln smooth bromegrass with progressively later harvest dates (W. G. Monson, unpublished data).

The cutting schedule for alfalfa must be balanced between nutrient content, forage yield, and stand persistence. In the northern U.S., cutting schedules of more than a 35 to 45-day regrowth interval will reduce nutrient digestibility and CP levels. This raises the question of whether alfalfa should be cut according to calendar date or growth stage. Since environment fluctuates from year to year, plants may reach certain stages of development at different times. In the northern U.S., harvesting according to growth stage seems to result in the least variation in feeding value. Harvesting at the one-tenth bloom stage is generally accepted as a compromise of quality and quantity in this region. Further south, plant size and lower leaf drop has been a more reliable indicator.

Of the perennial cool season grasses, perennial ryegrass maintains higher digestibility than other species (Smetham, 1973). In contrast, timothy, smooth bromegrass, orchardgrass, and crested wheatgrass [Agropyron cristatum (L.) Gaertn.] show a marked decline in digestibility with maturity. This difference is illustrated by the declining digestibility of smooth bromegrass in New York as forage was harvested at later dates (Fig. 6.12). Although dry forage yield increased with maturity, the yield of digestible dry matter declined when the percent digestibility dropped below 55%. Crude protein content also declined sharply with maturity.

Two range grasses also show a decline in digestibility with maturity, but junegrass [Koeleria cristata (L.) Pers.] maintains a higher quality than does crested wheatgrass (Fig. 6.13). Red clover, like alfalfa, declines in digestibility with maturity, but white clover maintains high digestibility for most of the year because only leaves and petioles are cut or grazed.

Cool season annual species, such as cereals, annual ryegrass, and annual clovers normally retain high digestible energy and protein throughout the winter and early spring growing season until bloom

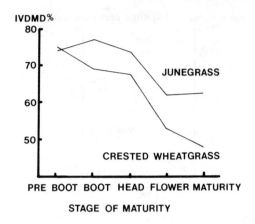

Fig. 6.13—Changes in in vitro dry matter digestibility (IVDMD) of two range grass species with advancing maturity (Raleigh, 1973).

stage. Digestibility and crude protein in ryegrass declines a little with maturity, while quality of cereal forage declines more steeply. Ryegrass and cereal forage are high in soluble sugars, a fact partially accounting for the high daily gains of growing animals on these pastures. Digestibility of annual clovers vary. These clovers typically decline rapidly in digestibility with the onset of flowering. Arrowleaf clover is an example; even though its digestibility declines, the rate is slower than that of crimson clover. Stems of crimson clover are less digestible than arrowleaf clover in early bloom stage, a result of more digestible cell walls in the latter species (Hoveland et al., 1970).

Nutritive quality of tropical perennial grasses is greatly influenced by stage of maturity at harvest. High light intensity, warm temperatures, and adequate moisture encourage rapid growth and maturation. Forage quality of bermudagrass, bahiagrass, guineagrass (*Pancium maximum* L.), and digitgrass decline rapidly with increasing maturity. With increasing age, Coastal bermudagrass forage declines in CP and digestibility (Fig. 6.14). Reduced forage quality results in lowered gains, as shown by dairy heifers in Georgia fed 4, 8, and 13-week old Coastal bermudagrass hay without supplements (Table 6.6). Although the reduction in digestibility may seem small in more mature hay, the daily gains are drastically reduced. The low protein and digestibility often attributed to forage of tropical (compared to temperate) species, can to a large extent, be improved by better management.

The decline in CP content with advancing maturity is generally more rapid with tropical than with temperate grass species. With both however, this decrease is a result of more stem and less leaf tissue at maturity. Seasonal variation in CP content is pronounced, with contents of 10 to over 15% at the beginning of the rainy season, declining to 5% or less during the winter dry season. Kikuyugrass maintains a

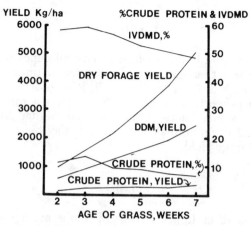

Fig. 6.14—Yield, in vitro dry matter digestibility (IVDMD), and crude protein content of Coastal bermudagrass at progressively later stages of development (W. G. Monson, unpublished data).

considerably higher CP content as it matures than does napiergrass (*Pennisetum purpureum* Schumach), guineagrass, molassesgrass (*Melinus minutiflora* Beau V.), digitgrass, or bermudagrass.

Advancing maturity also affects mineral composition. Potassium content decreases with maturity, generally more rapidly than Ca. Mature grass hay can be deficient in P for animal requirements.

Tropical legumes generally maintain higher levels of CP and DE than do warm season perennial grasses over the growing season. Even during the usual winter dry season in tropical areas, when digestibility and protein of grasses deteriorates, legumes provide higher quality forage.

It is relatively easy to maintain satisfactory nutrients in hay by harvesting at the proper time. However, managing to maintain high quality forage in a perennial grass pasture is more difficult and generally requires cross-fencing to allow frequent rotation of animals or close continuous grazing with supplemental feeding during periods of limited forage growth. Lax, continuous grazing, with a large accumulation of forage, may reduce digestibility (and sometimes protein) and lower animal performance.

Table 6.6—Effects of age on Coastal bermudagrass hay on performance of dairy heifers.†

Age of grass when cut	Crude protein	Crude fiber	Digest-ibility	Intake/ day		Average daily gain	
weeks		%		kg	lb	kg	lb
4	15.8	28.6	55	5.4	11.9	0.54	1.2
8	6.3	30.4	49	4.0	8.8	0.40	0.9
13	5.3	33.8	45	4.3	9.5	0	0.0

† From McCullough and Burton (1962).

Fertilization

Fertilization and liming of forage may influence nutritive quality in two ways. First, there may be a direct effect in uptake of a particular element such as N or P. Secondly, application of lime or fertilizer may influence the proportion of legume in a grass-legume sward. Since legumes are usually more digestible and higher in protein than grasses, the effect on animal performance may be more marked than the effect on forage yield.

Lime

Soil acidity often limits growth of temperate legumes more than grasses. Thus, liming to increase soil pH may favor growth of legumes which are highly digestible, protein-rich, and have high intake. As for direct effect on individual forage species, lime may increase digestibility and intake of grasses (Rees and Minson, 1976). Calcium, Mg, and P concentration are normally increased by liming acid soil. Overliming may reduce plant Mg concentration. Liming may increase micronutrient composition of forage plants by addition of Mn, Zn, Co, Cu, B, and Mo as impurities or alter the availability of these elements through changes in soil pH.

Nitrogen

The major effects of applying N fertilizer to grass are improved forage yield and CP concentration (Prine and Burton, 1956). Addition of N early in the growing season results in a rapid rise in CP but later in the year the effect is less. Generally, the levels of N needed to increase CP content of tropical grasses is less than that needed for temperate grasses. At the same stage of growth, N fertilization has little or no effect on digestibility (Webster et al., 1965). Thus, N increases carrying capacity and production per hectare, but outputs per animal are not generally improved.

Some undesirable changes in forage also may occur from N fertilization, such as elimination of legumes from a mixed grass-legume sward. These changes indirectly affect the uptake of minerals in forage available and reduce intake potential. High rates of N fertilizer can give a desirable high level of CP in grasses, sometimes 30% or more, much of this in the form of non-protein nitrogenous material that ruminants can utilize. However, high N fertilizer rates may also cause free nitrate N in excess of 0.4%, which can cause death. Rye and tall fescue accumulate more nitrate than bermudagrass (Hojjati et al., 1972). Maximum nitrate accumulation in tall fescue and Kentucky bluegrass occurs within 14 days after fertilization (Hojjati et al., 1973). Tall fescue retains high levels of nitrate longer than does Kentucky bluegrass.

Other adverse effects on N fertilization may be an increase in alkaloid content of reed canarygrass and HCN in sorghum and sudangrass. High rates of broiler house litter, containing high N, can cause hard fat accumulation and strangulation of the intestinal tracts of cattle.

Phosphorus

The major effect of P fertilization is to increase forage yield. Thus, application of P may have its greatest effect on the amount of animal product produced per land area rather than on gains per animal.

When fertilized to obtain satisfactory forage yields, P content is usually adequate for ruminant animals (above 0.2%). As mentioned previously, P deficiency has been a problem on sandy leached Florida soils and on some high pH soils in the western U.S. Cattle are more susceptible than sheep to P deficiency. Legumes show a greater response, in terms of P concentration, to fertilization than do grasses. On many soils, fertilizer application may have little effect on forage P content unless there is a severe deficiency.

Potassium

Although ruminant animals will not develop K deficiency on a forage diet, this element may have an important deleterious effect on forage quality of temperate grasses. Heavy K fertilization increases the content of this element in the forage while decreasing relative plant levels of Na, Ca, and Mg. These minerals may be marginal for needs of the grazing ruminant animal. Reduced Mg may result in grass tetany (Grunes, 1973).

Application of K fertilizer has little effect on utilization of forages. Animal production responses result primarily from 1) increased forage yield per hectare and 2) encouraging more clover or alfalfa in a mixed grass-legume sward.

Magnesium

Magnesium is normally applied to forages as dolomitic limestone. This application is adequate for pasture growth but may not supply sufficient Mg in the forage to prevent grass tetany. Normally, 0.2 to 0.25% Mg in forage is required to control the hypomagnesemia syndrome. Sometimes, grass Mg levels may drop to 0.10 to 0.15%. Fertilization of pastures has not been a dependable method to supply adequate Mg to livestock during the tetany-prone season of late winter and early spring.

Sulfur

It appears that if sufficient S is present for good forage growth, the S needs of the animal will be supplied. However, the greater use of high analysis fertilizers which contain little S has reduced forage yield on some soils. Although animals perform better on S-fertilized alfalfa hay, the primary effect seems to be improved protein content rather than S.

Micronutrients

On most soils, micronutrients are not a limiting factor in forage production. Boron is commonly added for alfalfa production. With higher rates of N, P, and K, which greatly increase forage growth, it can be expected that micronutrients will be depleted and more problems will appear in the future. Only a few elements, such as Co, Cu, and Se reduce animal productivity on all-forage diets. Severe deficiency occurs in Florida and Australia, but it can be overcome with fertilization or supplemental feeding of the required micronutrients.

Burning

Burning of pastures has long been regarded by cattlemen as a way to improve forage quality. Burning is commonly done on tropical grass pastures. Fire stimulates early growth of new leaf tissue, which has high digestibility and CP, and destroys the old low quality residue. In addition, burning may increase mineral availability near the soil surface, obtained from minerals in standing dead residue.

Early spring burning of Coastal bermudagrass sod is a recommended practice for control of the 2-lined spittlebug (*Prosopia bicincta* Say), plant diseases, and winter weeds. In a 6-year study at Tifton, Coastal bermudagrass burned about 1 March yielded an average of 1,000 kg/ha (892 lb/acre) more than the unburned sod (Monson et al., 1974). We found no effect on digestibility or CP content.

Pests

Diseases and insects may reduce the digestibility and CP of many forages by causing loss of leaves. This reduction in the leaf-stem ratio is particularly apparent in legumes such as alfalfa. In grasses, such as tall fescue, leaf rust may reduce soluble carbohydrates so that CP content is increased slightly. In orchardgrass, rust infection may reduce digestibility (Carlson, 1974).

Weeds may indirectly reduce forage quality in pastures or hay fields. Many weeds such as broomsedge (*Andropogon virginicus* L.)

Table 6.7—In vitro digestible dry matter (IVDMD) and crude protein content of weed species in Minnesota, 3-year average.†

Category	Species	IVDMD	Crude protein
		\%---------	
Palatable grasses	Yellow foxtail (*Setaria glauca*)	74	22
	Barnyardgrass (*Echinchloa crusgalli*)	79	22
	Green foxtail (*Setaria viridus*)	75	21
Unpalatable grass	Giant foxtail (*Setaria faberi*)	72	21
Palatable forbs	Redroot pigweed (*Amaranthus retroflexus*)	79	24
	Pennsylvania smartweed (*Polygonum pennsylvanicum*)	62	23
	Common lambsquarter (*Chenopodium album*)	71	26
Unpalatable forbs	Wild mustard (*Brassica kaber*)	69	20
	Giant ragweed (*Ambrosia trifida*)	72	22
	Common cocklebur (*Xanthium pennsylvanicum*)	77	24

† From Marten and Anderson (1975).

have a lower nutritive quality than the desired grass or legumes. Thus, as low-quality weeds become a significant population in the field they dilute the overall quality of forage available. Some weeds may be poisonous. Many weed species have high nutritive quality and can be valuable even though they have other undesirable characteristics as forages (Table 6.7). Many annual weed species do not decrease the nutritive value of hay or pasture if utilized at relatively early stages of maturity (Marten and Andersen, 1975). Other relatively high quality weeds such as smooth crabgrass [*Digitaria sanguinalis* (L.) Scop.] and little barley (*Hordeum pusillum* Nutt). can be valuable forage plants in the vegetative stage.

SUMMARY

Essential to forage quality are DE intake potential, protein, and minerals. These components are influenced both by genetic and environmental factors.

The genetic potential of forage species and genotypes within species differ considerably with respect to available energy and protein. A few generalizations can be made. Legume species are sometimes more digestible and often contain more protein than grasses. Legumes almost always have higher intake potential than grasses. Cool season grasses generally have higher nutritive value than tropical grasses. We note many exceptions to these generalizations, often caused by anti-quality components, such as alkaloids, tannins, saponins, cyanogenic glycosides, estrogens, coumarin, and mineral imbalances. Plant breeders have been able, with modern screening methods, to develop cultivars with higher nutritive quality. Great potential exists for further improvement in forage quality by plant breeding.

Environmental factors may greatly modify the genetic potential of forage plant quality. Climatic factors such as temperature, light, soil moisture, and soil oxygen influence nutritive quality. One of the most important factors affecting digestibility and protein content is stage of maturity. Age of the leaves and stems at time of cutting or grazing as well as season of the year affect quality components. Digestibility and protein both decline with age, although legumes generally retain their quality longer than grasses. Fertilization and liming influence quality directly via uptake of a nutrient and indirectly by changing the proportion of legume to grass in a sward. Lime and fertilizer nutrients do not directly affect digestibility. Liming may favor growth of legumes and uptake of Ca, Mg, and P by both grasses and legumes. The major effect of N, aside from increasing yield, is to improve protein content. Application of other plant nutrients mainly influence forage yield, although P and trace elements may be increased. Magnesium fertilization is normally inadequate to control grass tetany. Burning generally improves early spring forage quality. Diseases, insects, and weeds may either increase or decrease quality, depending on species and their stage of maturity.

LITERATURE CITED

Akin, D. E., F. E. Barton, II, Henry Amos, and Donald Burdick. 1973. Rumen microbial degradation of grass tissue revealed by scanning electron microscopy. Agron. J. 65:825–828.

Barnes, R. F., and C. H. Gordon. 1972. Feeding value and on-farm feeding. Agronomy 15:601–630. Am. Soc. Agron., Madison, WI.

————, and D. L. Gustine. 1973. Allelochemistry and forage crops. p. 1–13. In A. G. Matches (ed.) Anti-quality components of forages. Special Publ. No. 4. Crop Sci. Soc. Am., Madison, WI.

Blaser, R. E., H. T. Bryant, R. C. Hammes, Jr., R. L. Boman, J. P. Fontenot, and C. E. Polan. 1969. Managing forages for animal production-major research findings in detail. In Managing forages for animal production. Virginia Agric. Exp. Stn. Res. Bull. 45.

Burns, J. C., R. F. Barnes, W. F. Wedin, C. L. Rhykerd, C. H. Noller. 1970. Nutritional characteristics of forage sorghum and sudangrass after frost. Agron. J. 62:348–350.

Burton, G. W., R. H. Hart, and R. S. Lowrey. 1967. Improving forage quality in bermudagrass by breeding. Crop Sci. 7:329–332.

————, and W. G. Monson. 1972. Inheritance of dry matter digestibility in bermudagrass, Cynodon dactylon (L.) Pers. Crop Sci. 12:375–378.

————, ————, J. C. Johnson, Jr., R. S. Lowrey, H. D. Chapman, W. H. Marchant. 1969. Effect of the d₂ dwarf gene on the forage yield and quality of pearl millet. Agron. J. 61:607–612.

Bush, L., and R. C. Buckner. 1973. Tall fescue toxicity. p. 99–112. In A. G. Matches (ed.) Anti-quality components of forages. Special Publ. No. 4. Crop Sci. Soc. Am., Madison, WI.

————, R. C. Streeter, and R. C. Buckner. 1970. Perloline inhibition of in vitro ruminal cellulose digestion. Crop Sci. 10:108–109.

Carlson, I. T. 1974. Correlations involving in vitro dry matter digestibility of Dactylis glomerata L. and Phalaris arundinacea L. p. 732–738. In Proc. XII Int. Grassland Crop., Moscow, USSR. Vol. III.

Cooper, J. P. 1973. Genetic variation in herbage constituents. p. 379–417. *In* G. W. Butler and R. W. Bailey (eds.) Chemistry and biochemistry of herbage. Vol. 2. Academic Press, NY.

————, J. M. A. Tilley, W. F. Raymond, and R. A. Terry. 1963. Selection for digestibility in herbage grasses. Nature 195:1276–1277.

Deinum, B. 1966. Influence of some climatological factors on the chemical composition and feeding value of herbage. p. 415–418. *In* Proc. X Intern. Grassland Congr., Helsinki, Finland.

————, A. J. H. Van Es, and P. J. Van Soest. 1968. Climate, nitrogen and grass. II. The influence of light intensity, temperature and nitrogen on in vivo digestibility of grass and the prediction of these effects from some chemical procedures. Neth. J. Agric. Sci. 16:217–223.

Donnelly, E. D., and W. B. Anthony. 1969. Relationship of tannin dry matter digestibility and crude protein in *Sericea lespedeza*. Crop Sci. 9:361–362.

————, and ————. 1970. Effect of genotype and tannin on dry matter digestibility in *Sericea lespedeza*. Crop Sci. 10:200–202.

Elkins, C. B., R. L. Haaland, C. S. Hoveland, and W. A. Griffey. 1978. Grass tetany potential of tall fescue as affected by soil O_2. Agron. J. 70:309–311.

Gengenbach, B. G., and D. A. Miller. 1972. Variation and heritability of protein concentration in various alfalfa plant parts. Crop Sci. 12:767–769.

Gil, H. C., R. L. Davis, and R. F. Barnes. 1967. Inheritance of in vitro digestibility and associated characteristics in *Medicago sativa*. Crop Sci. 7:19–21.

Grunes, D. L. 1973. Grass tetany of cattle and sheep. *In* A. G. Matches (ed.) Anti-quality components of forages. Special Publ. No. 4. Crop Sci. Soc. Am., Madison, WI.

Hanna, W. W., W. G. Monson, and G. W. Burton. 1973. Histological examination of fresh forage leaves after in vitro digestion. Crop Sci. 13:98–102.

Hanson, C. H., M. W. Pederson, B. Berrang, M. E. Wall, and K. H. Davis, Jr. 1973. The saponins in alfalfa cultivars. p. 33–52. *In* A. G. Matches (ed.) Anti-quality components of forages. Special Publ. No. 4. Crop Sci. Soc. Am., Madison, WI.

Harris, R. R., E. M. Evans, J. K. Boseck, and W. B. Webster. 1972. Fescue, orchardgrass, and Coastal bermudagrass for yearling beef steers. Auburn Univ. (AL) Agric. Exp. Stn. Bull. 432.

Heinrichs, D. H., and J. E. Troelson. 1965. Variability of chemical constituents in an alfalfa population. Can. J. Plant Sci. 45:405–412.

Hojjati, S. M., T. H. Taylor, and W. C. Templeton, Jr. 1972. Nitrate accumulation in rye, tall fescue, and bermudagrass as affected by nitrogen fertilization. Agron. J. 6:624–627.

————, W. C. Templeton, Jr., T. H. Taylor, H. E. McKean, and J. Byars. 1973. Postfertilization changes in Kentucky bluegrass and tall fescue herbage. Agron. J. 65:880–883.

Holter, J. A., and J. T. Reid. 1959. Relationship between the concentrations of crude protein and apparently digestible protein in forages. J. Animal Sci. 18:1339.

Hoveland, C. S., and W. B. Anthony, J. G. Starling. 1978. Beef cow-calf performance on Coastal bermudagrass overseeded with winter annual clovers and grasses. Agron. J. 70:418–420.

————, E. L. Carden, W. B. Anthony, and J. P. Cunningham. 1970. Management effects on forage production and digestibility of Yuchi arrowleaf clover (*Trifolium vesiculosum* Savi). Agron. J. 62:115-116.

————, R. F. McCormick, and W. B. Anthony. 1972. Productivity and forage quality of Yuchi arrowleaf clover. Agron. J. 64:552-555.

Johnson, J. C., Jr., R. S. Lowrey, W. G. Monson, and G. W. Burton. 1968. Influence of the dwarf characteristic on composition and feeding value of near-isogenic pearl millets. J. Dairy Sci. 51:1423-1425.

————, W. G. Monson, G. W. Burton, and W. C. McCormick. 1976. Performance of dairy heifers grazing pastures of either Gahi-1, Gahi-3, or Tifleaf-1 millet. J. Dairy Sci. 59:19.

Jones, W. T., and J. W. Lyttleton. 1971. Bloat in cattle. XXXIV. A survey of legume forages that do and do not produce bloat. N.Z. J. Agric. Res. 14: 101-107.

Julen, Gosta, and Agneta Lager. 1966. Use of the in vitro digestibility test in plant breeding. p. 652-657. *In* Proc. X Intern. Grassland Congr., Helsinki, Finland.

Lowrey, R. S., G. W. Burton, J. C. Johnson, Jr., W. H. Marchant, and W. C. McCormick. 1968. In vivo studies with Coastcross-1 and other bermudas. Georgia Agric. Exp. Stn. Res. Bull. 55.

McCullough, M. E., and G. W. Burton. 1962. Quality in Coastal bermudagrass hay. Georgia Agric. Res. 4:4-5.

Marten, G. C. 1973. Alkaloids in reed canarygrass. p. 15-31. *In* A. G. Matches (ed.) Anti-quality components of forages. Special Publ. No. 4. Crop Sci. Soc. Am., Madison, WI.

————, and R. N. Anderson. 1975. Forage nutritive value and palatability of 12 common annual weeds. Crop Sci. 15:821-827.

Milford, R. 1967. Nutritive values and chemical composition of seven tropical legumes and lucerne grown in sub-tropical south-eastern Queensland. Aust. J. Exp. Agric. Anim. Husb. 7:540-545.

————, and D. J. Minson. 1963. Intake of tropical pasture species. p. 815-822. *In* Proc. IX Intl. Grassland Congr. (Sao Paulo, Brazil).

Minson, D. J., C. E. Harris, W. F. Raymond, and R. Milford. 1964. The digestibility and voluntary intake of S-22 and H-1 ryegrass, S-170 tall fescue, S-48 timothy, S-215 meadow-fescue and germinal cocksfoot. J. Brit. Grassland Soc. 19:298-305.

————, and M. N. McLeod. 1970. The digestibility of temperate and tropical grasses. p. 719-722. *In* Proc. XI Intern. Grassland Congr., Surfers Paradise, Queensland, Australia.

Monson, W. G., G. W. Burton, E. J. Williams, and J. L. Butler. 1974. Effects of burning on soil temperate and yield of Coastal bermudagrass. Agron. J. 66:212-214.

————, J. B. Powell, and G. W. Burton. 1972. Digestion of fresh forage in rumen fluid. Agron. J. 64:231-233.

Morey, D. D., M. E. Walker, W. H. Marchant, and R. S. Lowery. 1969. Small grain forage production and quality as influenced by rates of nitrogen. Univ. of Georgia, College of Agric. Exp. Stn. Res. Bull. 70.

Muller, L. D., R. F. Bauman, and F. V. Colenbrender. 1971. Variations in lignin and other structural components of brown midrib mutants of maize. Crop Sci. 11:413-415.

Prine, G. M., and G. W. Burton. 1956. The effects of nitrogen rate and clipping frequency upon the yield, protein and content and certain morphological characteristics of Coastal bermudagrass [*Cynodon dactylon* (L.) Pers.] Agron. J. 48:296-301.

Raleigh, R. J. 1973. Range forage and animal nutrition. p. 17-30. *In* M. J. Wright (ed.) Range research and range problems. Spec. Publ. No. 3. Crop Sci. Soc. Am., Madison, WI.

Raymond, w. F. 1969. The nutritive value of forage crops. Adv. Agron. 21:2-97.

Rees, M. C., and D. J. Minson. 1976. Fertilizer calcium as a factor affecting the voluntary intake, digestibility, and retention time of pangola grass (*Digitaria decumbens*) by sheep. Br. J. Nutr. 36:179-187.

Reid, R. L., and G. A. Jung. 1973. Forage-animal stresses. p. 639-653. *In* M. E. Heath, D. S. Metcalfe, and F. R. Barnes (ed.) Forages. Iowa State Univ. Press, Ames, IA.

Roth, L. S., G. C. Marten, W. A. Compton, and D. D. Stuthman. 1970. Genetic variation of quality traits in maize (*Zea mays* L.) forage. Crop Sci. 10:365-367.

Simon, A. B., and G. C. Marten. 1971. Relationship of indole alkaloids to palatability of *Phalaris arundinacea* L. Agron. J. 63:915-919.

Sleper, D. A., P. N. Drolsom, and N. A. Jorgensen. 1973. Breeding for improved dry matter digestibility in smooth bromegrass (*Bromus inermis* Leyss). Crop Sci. 13:556-558.

Smetham, M. L. 1973. Grazing management. p. 179-228. *In* R. H. M. Langer (ed.) Pastures and pasture plants. A. H. and A. W. Reed, Wellington, N.Z.

Smith, D. 1969. Influence of temperature on the yield and chemical composition of 'Vernal' alfalfa at first flower. Agron. J. 61:470-472.

————. 1970. Influence of cool and warm temperature reversal at inflorescence emergence on yield and chemical composition of timothy and bromegrass at anthesis. p. 510-514. *In* Proc. 11th Intern. Grassland Congr., Surfers Paradise, Queensland, Australia.

Smith, W. H., V. L. Lechtenberg, D. C. Petritz, and D. G. Hawkins. 1975. Cows grazing orchard, fescue, or fescue-legume. J. Animal Sci. 41:339-340.

Stuthman, D. D., and G. C. Marten. 1972. Genetic variation in yield and qualith of oat forage. Crop Sci. 12:831-833.

Utley, P. R., H. D. Chapman, W. G. Monson, W. H. Marchant, and W. C. McCormick. 1974. Coastcross-1 bermudagrass, Coastal bermudagrass, and Pensacola bahiagrass as summer pasture for steers. J. Animal Sci. 38:490-495.

Vough, L. R., and G. C. Marten. 1971. Influence of soil moisture and ambient temperature on yield and quality of alfalfa forage. agron. J. 63:40-42.

Webster, J. E., J. W. Hogan, and W. C. Elder. 1965. Effect of rate of ammonium nitrate fertilization and time of cutting upon selected chemical components and the in vitro digestion of bermudagrass forage. Agron. J. 57:323-325.

SUGGESTED READING

Harrison, C. M. (ed.). Forage economics-quality. Spec. Publ. 13. Am. Soc.
    Agron., Madison, WI.
Hoveland, C. S. (ed.). 1976. Biological N fixation in forage-livestock systems.
    Spec. Publ. 28. Am. Soc. Agron., Madison, WI.
Matches, A. G. (ed.). 1973. Anti-quality components of forages. Spec. Publ. 4.
    Crop Sci. Soc. Am., Madison, WI.
Mays, D. A. (ed.). 1974. Forage fertilization. Am. Soc. Agron., Madison, WI.

Chapter 7

Genetic and Environmental Effects on Quality of Fiber Crops

B. A. WADDLE

University of Arkansas
Fayetteville, Arkansas

Crops are grown for their fiber either as a direct plant product as in the case of seed hairs in cotton (*Gossypium* spp.) or as a processed plant product. The latter fiber includes structural or bast fibers located in stems, leaves, or petioles and must be stripped from companion cells by various processes.

In crops grown for structural fibers the whole plant or leaf is harvested by cutting or pulling. Fibers are removed by decortication. The decortication process varies in small detail from crop to crop. Principal examples of such internationally grown crops are henequen (*Agave fourcroydes* Lem.), sisal (*Agave sisalana* Perr.), abaca (*Musa textilis* Nee.), jute (*Corchorus* spp.), hemp (*Cannabis sativa* L.), and flax (*Linum usitatissimum* L.). Another fiber crop, kenaf (*Hibiscus cannabinus* L.), may be included in this group because of its potential for paper production. Structural fibers consist mainly of cellulose.

COTTON

Cotton fibers are exposed at maturity. At harvest the seed and attached fibers are removed from the plant. A process called ginning removes the fiber from the seed. This ginned fiber, almost pure cellulose, is ready to be spun into yarn and then woven into fabric or used otherwise as described in chapter 2.

Seed and seed hairs of cotton develop inside a capsule or boll and require 50 to 80 days to mature. A single plant has few to many bolls (Fig. 7.1). The period of boll set or fruiting period for a given field of cotton extends from 2 to 4 months. Quality of the cotton fiber responds to genetic and to environmental factors during this more or less extended fruiting period.

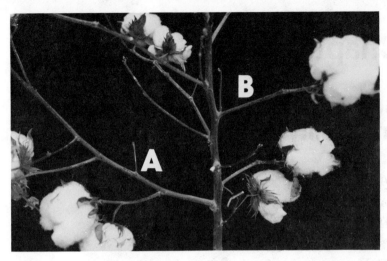

Fig. 7.1—Mature cotton plant showing vegetative branch (A) with several fruiting branches and (B) a fruiting branch arising from the main stem. (Photo courtesy of R. E. Stevenson, Auburn Univ. Agric. Exp. Stn., Auburn, AL).

Fiber quality in cotton has many dimensions, as described in detail in chapter 2. Grade and staple length are gross measurements of fiber quality.

Grade in ginned cotton represents the color, "leaf," or trash content, and gin preparation of the sample being evaluated for quality. Of these, color and trash content are the more common grade determinants. A color change from normal white in cotton reflects some degree of field weathering. Color changes may also be associated with insect and disease injuries in the field and with soil splattered onto the open bolls of cotton during rainstorms. Stains by juices from plants crushed in the harvest process, and especially if followed by high drying temperatures imposed in the ginning process, may cause further color change. A color change also may occur if the harvested product contains more than 12% moisture by weight and is held for an indefinite period prior to ginning. Prematurely opened cotton bolls with immature fibers will deteriorate in color more rapidly than will mature cotton. Trash is incorporated into cotton during the harvest operation and is removed in ginning. Leaf, bract, stem, and boll fragments from the cotton plant and resident weed pest fragments are sources of cotton trash.

Fiber length, as determined by standardized cotton stapling procedures, is more than an estimate of absolute length. Official government classers in the USA report the length in 32nds of an inch, and this length prevails in the international marketing system both in countries using the English and in those using the metric systems. The USSR uses the metric system for staple length. Length classification is largely a function of average fiber length (upper half mean), but includes such intangibles as "drag" or friction, fiber maturity, and fiber length uniformity. Fiber length can never be increased beyond that present when the cotton is harvested. Ginning practices, however, can shorten the average fiber length by cutting or breaking individual fibers.

Both grade and staple length are highly subjective indices of value. Cotton as fiber generally moves in trade channels, however, on the basis of grade and staple length. Wide variations in fiber quality in cotton exist even for such gross measurements as grade and staple. Essentially all of this variation is the product of specific genotypes grown, of the total environment including harvesting and ginning, and of genotype-environment interactions.

Plant Growth Roles in Quality of Fiber

Genetic and environmental effects on cotton fiber quality reflect the unique fruiting pattern in cotton. As grown commercially, each cotton plant essentially is one big flowering raceme (Fig. 7.1). The first fruit develops on the first fruiting branch which departs from the main stem anywhere between the sixth and tenth node. Flower buds, called "squares," are visible 2 to 3 weeks before flowering. The flower is large and showy. The corolla is white the day of anthesis and then turns red the second day. The corolla dries and falls off to expose the young boll in 3 to 5 days. White flowers can be found at approximately 3-day intervals at the first nodal position on successive branches up the main stem. This stage is often referred to as the vertical flowering interval (VFI) and may range from 1 to 7 days on a single plant's main stem. White flowers appear at approximately 6-day intervals at successive nodal positions on a single fruiting branch. This stage is often referred to as the horizontal flower interval (HFI) and may range from 3 to 10 days among fruiting branches on a single plant. The main stem of a cotton plant may have as many as 30 nodes with fruiting branch potentials. A single fruiting branch may support as many as six or seven fruiting positions. Flowering, once initiated, continues over a period of 4 to 10 weeks or longer. All flowers do not result in bolls that mature. Approximately 60 days are required for the period from flowering to open boll when the seed-borne fibers are mature. First harvest normally is initiated when 70% of all the bolls are open. The harvested lint, therefore, will have been subjected to seasonal variations spreading over a period of 80 to 120 days.

The cotton fiber grows in length as a single cell extension from the seedcoat inside the bolls for a period of about 21 days after the white flower is visible. Secondary wall thickenings are initiated along the outer cell perimeter with cellulose being deposited in layers toward the center, simulating daily growth rings, for the next 30 days or until the boll opens. These processes are described in detail in chapter 2. During the fiber elongation period, water stress in the plant, temperatures below approximately 17 C (60 F), disease infection, or leaf damage from any source will slow the growth of these single cells and may reduce fiber length. The daily growth rings of deposited cellulose simulate

Table 7.1—Fiber properties of four cultivars grown in four major cotton regions of the
United States in 1972.†

Cultivar‡	Fiber length 2.5% span Region of growth§				Tensile strength# Region of growth				Maturity (micronaire)¶ Region of growth			
	A	B	C	D	A	B	C	D	A	B	C	D
	mm				mN				units			
Coker 310	30.7	30.7	30.5	30.5	358	371	358	363	4.50	4.60	4.69	4.97
Deltapine 16	29.7	30.0	29.7	29.7	333	352	338	335	4.52	4.65	4.60	4.91
Lockett 4789A	28.7	28.7	28.2	28.7	354	371	357	363	4.24	4.27	4.47	4.66
Acala 1517-70	29.5	30.2	30.5	30.5	429	447	442	442	3.85	4.06	4.34	4.37
Regional avg.	29.7	30.0	29.7	30.0	368	385	374	376	4.28	4.40	4.52	4.73

† Adapted from Ramey et al. (1975).
‡ Cultivars developed specifically for the Eastern, Delta, Plains, and Western Regions Respectively.
§ Averages of 7 to 9 locations in each region. A = Eastern, B = Delta, C = Plains, and D = Far West.
# Tensile strength is T-0 gauge as measured by the "Stelometer" in milliNewtons (mN) per tex where
tex is the linear density of fibers expressed as gram weight of 1.000 m of fiber.
¶ Micronaire is an air resistance instrument reflecting maturity in that low numbers within a culti-
var are less mature than high numbers.

daily records of plant activity. Bright sunny days approaching 30 C
day and 18 C night temperatures surrounding plants free of disease
and insects and growing in soil with adequate moisture and nutrients
give maximum fiber thickness. Such fibers are described as fully mature. Departures from this optimum may reduce width of daily growth
rings and subsequently reduce fiber thickness or maturity index, as reflected in lowered micronaire values, and, indirectly lengthen the boll
maturation period beyond the 60 days described above.

Genetic Control of Fiber Properties

Genetic differentials in quality (grade and staple length) become
evident when several cultivars are planted on the same date on the
same soil type and are harvested on the same date (Table 7.1). Cultivars of cotton designed to utilize a long and continuous growing season, generally have longer fibers than those designed to produce acceptable yields under a growing season reduced in length. For example, in the U.S. Plains cultivars generally produce shorter fibers
than those developed in the Delta. 'Lockett 4789 A,' developed
specifically for rapid fruiting to exploit a short growing season, produces shorter fibers in all regions than those cultivars developed to exploit a full growing season. Environmental differentials in fiber length
shown in regional averages, are less than cultivar differentials at any
location. The same general trend holds for the other basic fiber properties as emphasized in chapter 2.

Cotton cultivars have been developed which reflect grade advances, but a few have been released which resulted in grade declines.
Grade changes were associated with degree of leaf pubscence. Cultivars having leaves smoother than normal result in less trash in the
ginned lint and they give a slight but consistent grade improvement.

Those having leaves with increased pubescence give lower grades because of increased trash in the ginned lint. Characters other than leaf pubescence are under genetic control and have demonstrated a potential for unique grade enhancement. Among these are the rolled bract (Frego bract), reduced bract, and withering bract. Cotton breeders historically have worked in concert with those involved in spinning and processing the fiber. Quality parameters generally are sufficiently heritable to warrant early generation selection for desired combinations. As emphasized in chapter 2, the wide range of cotton fiber end-uses require varying combinations of fiber properties. The strength of cotton as a fiber in the textile market reflects its diversity of predictable and measurable fiber properties. New cultivars designed to meet new demands by the textile industry can be anticipated because of the current genetic control of quality factors.

Environmental Effects on Fiber Quality

Grade deterioration in cotton usually is associated with field weathering. Cotton fiber deteriorates in the field when the grower is unable to harvest the crop as it matures. Crop maturity may be out of step with recurring seasonal conditions or the grower is unable to retain control of crop maturity. Positive action can be taken to reduce grade losses.

Any area of cotton production has recurring climatic conditions. In many areas of the world where cotton is a commercial crop, the grower experiences two seasons, a wet period and a dry period. In many production areas of the USA, the probability of adverse weather during the harvest period can be calculated. For example, in the Mississippi Delta Region of Arkansas, weather records for 50 years have been examined and a most probable favorable harvest weather period has been established as 1 to 25 October. A most probable period for field weathering has been established as 15 August to 25 September. The challenge for the grower here is to synchronize his crop with the favorable harvest period. This timing is done with his choice among cultivars, dates of planting, levels of fertilization, and by timing of defoliation. The same synchronization can be established for all cotton growing regions of the world.

Field weathering (color changes from white to gray, with or without the tinge or spots appearing in white grades) is more common with immature than mature fibers. Fiber maturity is reflected in fiber resistance to air flow as measured by micronaire and other air-flow instruments. A cultivar like 'Stoneville 7A,' as grown in the USA, has an average micronaire index ("mike") value of 4.8. If the mike index drops below 3.5, the fiber of this cultivar would be immature. Under better than average light and moisture conditions mike index of this cultivar will go above 5.0. Other cultivars have other genetic maturities for fiber (Table 7.2).

Table 7.2—Grade, staple, and fiber maturity (micronaire indices) of four cultivars grown 3 years at the same loction (Upper Delta Region, USA).†

Cultivar	Grade index‡			Steple length§			Micronaire indices#		
	1967	1968	1969	1967	1968	1969	1967	1968	1969
	——— no. ———			——— mN ———			——— units ———		
Stoneville 7A	83.0	94.0	97.8	274	279	286	3.3	4.8	4.8
Coker 413	81.0	92.8	95.5	270	279	286	3.1	4.0	4.1
Acala SJ-1	84.5	91.5	94.0	290	284	286	3.1	4.6	4.9
Bryant 79	84.5	94.0	97.0	278	279	286	2.9	4.1	4.2
Avg.	83.2	93.1	96.1	278	281	286	3.1	4.4	4.5

† Adapted from Lafferty et al. (1971).
‡ The higher the index, the better the quality as reflected in grade (Middling-White = 100).
§ Staple expressed in mN but based on U.S. government standards and classed by an official classer.
# Lower micronaire indices represent degree of immaturity of fibers harvested.

Fiber maturity may be interrupted by several factors. Prolonged water stress, root restrictions by hard pans, disease infection, nutrient deficiencies, misuse of herbicides and/or defoliants or desiccants, and extended periods of low temperatures are factors that may be associated with some degree of fiber immaturity. The most flagrant misuse of defoliants or desiccants is their premature use. Both terminate all leaf activity in cotton. All of these except low temperatures can be avoided by cotton growers. Advance knowledge of weathering predisposition can be used to assign field priorities in harvesting schedules. This management challenge is present wherever cotton is a commercial crop.

The impact of environment on fiber quality is considered for one area in Table 7.2. Grade differentials reveal that weathering was essentially the same for four cultivars grown in the Upper Delta Region of the USA in 1968 and 1969. Lower grades in 1967 were associated with lower micronaire indices. The fiber was immature because of lower than normal temperatures in September when daily growth rings of cellulose were being laid down. Staple length, which was presumed to be determined largely in late July and early August, was, in fact, hardly affected by the low September temperatures of 1967.

Interaction of Genetic and Environmental Factors

Environmental effects on grade and staple are not the same for all cultivars. Cultivars having genetic potentials for high micronaire indices do not have the weathering and subsequent grade loss potentials of their low mike counterparts. Bolls from low mike cultivars retain sugars and other non-used cellulosic food reserves on the fiber surface when the bolls open. The resulting microbial action accelerates color changes with any degree of field weathering.

Similarly, high mike cottons clean up better in the cleaning processes imposed before and after ginning. The increased use of mechanical harvesters in most of the world cotton-producing areas after World War II has been associated with increasing micronaire indices needed to assure trash-free fiber.

Table 7.3—Cultural practices that affect cotton fiber properties.†

Cultural practice	Effects on plant	Effect on fiber properties‡		
		Length	Strength	Maturity
Sub-soiling	Permits more extensive root system and lowers water stress, in pan-type soils	+	– or 0	+
Seed treatment	Gives more uniform stand and avoids skips	0 or +	0	– or 0
Skip row	Reduces water stress and extends fruiting period	+ or 0	0	+ or 0
Adding potash to to potash de-ficient soil	Avoids rust and premature defoliation	+	– or 0	+
Irrigation in rainbelt	Taller, fruiting period extended	+	– or 0	–
Too much nitrogen	Rank growth, bigger leaves and fruiting period extended	+	– or 0	–
Weed control with chemicals	Avoids moisture stress to slight extent	+ or 0	0	0
Defoliation 50% open	Normal leaf drop	0	0	– or 0
Defoliation 30% open	Fair leaf drop and regrowth	– or 0	0	–
Adding boron to a boron deficient soil	Improves fruit set	+	+	+

† Adapted from Anon. (1967).
‡ "+" = increase; "–" = to decrease; and "0" = no effect.

An exception to the two high mike generalities above is to be found in cultivars derived from primitive Hopi Indian cottons. These cultivars produce a fiber that is fully mature even though its mike can be considered as low. The fibers are more round in cross-section and resist weathering. They have the same trash cleanability as cultivars having significantly higher micronaire values.

Cultivars having what have been described as "high luster" fibers have a heavier than normal protective wax covering on individual fibers. This wax gives some degree of field weathering resistance.

Production Practices that Affect Cotton Fiber Quality

Cotton growers in all parts of the world have some control over fiber quality. This control is obtained in the choices of cultivar and cultural practices (Table 7.3). Predictable fiber quality effects can be given for any cultural practice. For example, choosing a genetically high mike cultivar for planting in an area known to have a shortened growing season will assure the grower of an acceptable fiber mironaire value for the cotton produced. Cultural practices have a direct affect on the cotton plant and indirectly affect such quality components as *length, strength,* and fiber *maturity* (mike). Any cultural practice that extends the effective fruiting period for the crop increases the probability of immature fibers having an accelerated field weathering risk. Irrigation in the rainbelt and excessive rates of applied N are examples of cultural practices that extend the effective fruiting period. Any cultural practice which maximizes fruit set during periods favoring fiber development should increase fiber length and maturity. Sub-soiling where needed and adding K or B to deficient soils are examples of such

cultural practices. Adding B to correct a B deficiency has been known
to increase fiber strength as a unique but puzzling response to a
specific cultural practice. Any cultural practice which affects leaf
activity should affect fiber properties, as illustrated by premature defoliation in Table 7.3. Defoliants, if properly applied, cause the leves to
drop from the plant and desiccants kill the leaves in place. Obviously,
premature use (only 30% open bolls) of either can result in drastic yield
reductions as well as quality reductions. Higher yielding cottons will
suffer greater losses.

Effects of Harvesting and Ginning on Fiber Quality

In the USA, seedcotton is mechanically harvested. The machines
used are either pickers or strippers. Mechanical pickers (Fig. 7.2) are
tractor-mounted, rotating spindles that entangle the seedcotton exposed in open bolls, removing it from the plant. Doffers strip the seedcotton from the spindles and an air conveyor delivers it into an overhead basket. More than one harvest can be and often is made of a
single field. Mechanical strippers (Fig. 7.3) are used once-over a field
and are of two types. The finger-type stripper is like a picket fence
pushed through the mature plants, and strips off open bolls. The other
type uses steel rollers or rotating brushes to strip the mature bolls
from the plant. Both kinds of strippers also strip off immature bolls
and even branches which are mixed into the harvested product. Both

Fig. 7.2—Two-row mechanical picker. Note unopened green bolls silhouetted against
wheels. These will open and be picked at a later date. (Photo courtesy of Deere & Co.).

Fig. 7.3—Self-propelled 2-row brush-type stripper with overhead basket on general use tractor. (Photop courtesy of Deere & Co.).

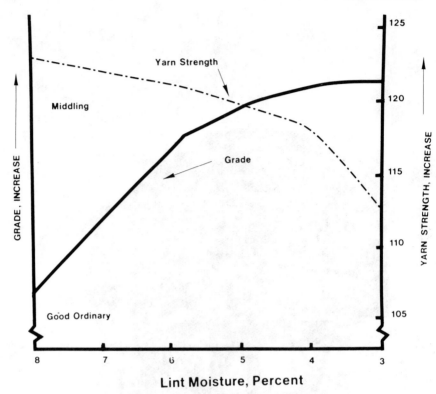

Fig. 7.4—Effects of gin drying on lint grade, yarn strength, and yarn appearance.

harvesting machines incorporate considerable trash into the unginned cotton, but the stripped product contains much more than picked cotton. Moisture control of seedcotton at the cotton gin is essential for optimum cleaning of machine-harvested cotton. Maximum trash removal from harvested seedcotton prior to ginning is obtained when seedcotton moisture is about 5% (Fig. 7.4). Removal of trash after ginning with lint cleaning machinery generally results in some grade improvement (Fig. 7.5). Final trash content, however, is largely a function of initial trash content.

Defoliation and/or desiccation prior to harvest generally reduces initial trash content of the harvested product. Harvested adjustments, plant densities, weeds, and cultivar planted affect initial trash content of the harvested product, generally in the order listed.

Mature seedcotton is harvested across fields from plants of varying maturity and hence varying in one or more quality parameters. The harvested product is a blended bulk when delivered to the gin. In ginning, the seedcotton is removed from its vehicle of transport and passed through drying and cleaning processes which tend to blend the cottons further. The dry and cleaned seedcotton is presented to the gin, either a saw or a roller type, almost as single seeds and attending seed hairs with no identify as to specific boll, plant, or place in the

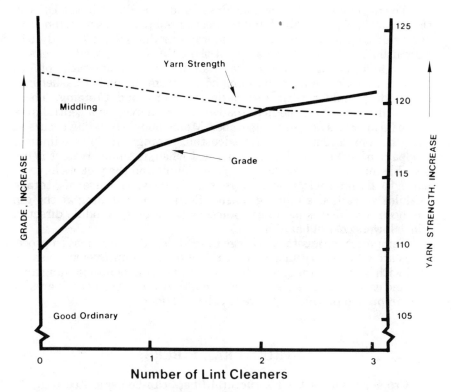

Fig. 7.5—Effect of lint cleaners on lint grade, yarn strength, and yarn appearance.

field. The single seeds are massed in some kind of roll against the ginning mechanism and more blending occurs. The freshly ginned lint is then cleaned and more blending occurs before packaging into bales.

The dynamic environment shaping the quality of fiber produced over a period of 80 to 120 days, therefore, is largely nuetralized by the blending action of harvesting and ginning. This may explain why the gross cotton quality measurements of grade and staple in the USA remain superior value determinants when compared to more sophisticated quality measurements showing fiber length, strength, maturity and their combined effort as indicated in yarn strength index.

Cotton fiber that enters trade channels before ginning, which happens in parts of Africa and in the USSR, may largely be harvested from localized areas or small fields having little plant to plant variation. Collecting centers in these areas accomplish the blending associated with harvesting and ginning in the USA. Unginned cotton is collected and stacked and held for ginning; some stacks in the USSR, for example, hold seed-cotton sufficient for as many as 5,000 bales of ginned lint. When these stacks are ginned, there is, in fact, such a high degree of blending that they achieve a high degree of uniformity, more so than with cottons of U.S. origin.

The development of new machinery and procedures in the USA in the early 1970's has stimulated interest in temporary seedcotton storage prior to ginning. Seedcotton storage modules in 1977 and 1978 were used in all production areas of the USA and in specific areas of Mexico and Central America. Free-standing seedcotton stacks of 5 to 15 bales each are built by dumping cotton from the picker basket into the module builder with automatic spreading and tamping. These stacks may be left on the field roadways, and moved to a central area of the farm or to a cooperating ginner. Much module building occurs at the gin from trailer loads of seedcotton. Storage modules eliminate stoppage of pickers during peak harvest when, in some areas, it is impossible for gins to clear their yards overnight and the pickers have no place to dump their loads. More significantly, the use of storage modules spreads the ginning season. Seedcotton handling at the gin has been modified sometimes to accept the storage module directly, which lowers cost of handling.

Temporary seedcotton storage in modules, if recommended procedures are followed, results in a degree of quality improvement associated with better cleaning with less artificial drying before ginning and, in some cases, less lint cleaning at the gin. All of these benefits result in an improved product delivered at the mill door.

STRUCTURAL FIBERS

Crops grown for their structural fibers can be separated roughly into hard and soft fiber groups. Hard fiber crops considered here are henequen, sisal, and abaca. Their fiber, obtained from leaves, is used primarily for cordage, ropes and twines. Soft fiber crops are jute, hemp, flax, and kenaf, and these are generally utilized in woven form or as pulp in the case of kenaf. All are structural fibers which are subjected to some form of decortication. They represent many countries as to origin in commercial trade and hence are subjected to an array of grading or quality measuring procedures as covered in chapter 2.

Factors Affecting grade in Structural Fibers

Extracting structural fibers from plant tissues is a variable process and the process used is their greatest single grade determinant (Table 7.4). The leaf fibers (sisal and henequen), if not cleaned properly, undergo rapid deterioration in grade factors of color and tensile strength. Abaca, the leaf sheath fiber, extracted by scraping, is sensitive to sunlight damage if exposure is extended beyond a narrow range for proper grade maintenance. The phloem or soft fibers are retted prior to fiber extraction. Retting, exposure to water, is a controlled fermentation process that releases the structural fibers to make them extractable. Color and tensile strength, major grade and end-use determinants, reflect retting skills regardless of fiber involved.

Table 7.4—Gross summarization of the relative importance of genetic and environmental factors in determining the quality of structural fiber crops.

	Relative rankings of quality determinants				
Fiber crops	Decortication process	Cultivar planted	Seasonal conditions	Weed control	Soil amendments
Henequen	1†	5	4	2	3
Sisal	1	5	4	2	3
Abaca	1	2	3	5	4
Jute	1	2	3	5	4
Hemp	1	3	2	5	4
Flax	1	2	3	4	5
Kenaf	2	1	5	3	4

† Relative order of importance of henequen fiber quality determinants, as an example, are decortication process, weed control, soil amendments, seasonal conditions, and cultivar planted.

Sisal and henequen crops are from vegetatively propagated plants that have not been subjected to breeding for improvement. Cultivars within species do not exist as commercial entities and, hence, have no influence on quality. Age of plants, cultural practices, especially weed control, and fertilization influence fiber length. Longer fibers come from better growing plants. Abaca cultivars have emerged and indirectly affect quality by influencing maturity and disease control or suppression. A product of tropical "milpa" agriculture, abaca is classed as a poor crop and little, if any, fertilizer is used in a plantation. Milpa agriculture refers to the clearing of jungle vegetation into small patches 1 to 3 ha in size. Crops are planted in these small patches. After a few crops, the native fertility is exhausted. The old patch is abandoned and a new milpa is cleared. The role of cultural practices in determining quality, therefore, is minimal.

Many cultivars of jute are utilized both in *C. capsularis*, which grows well in both lowlands and uplands, and in *C. olitorius*, which is grown only on upland soils. Cultivars differ in insect and disease resistance, maturity, and height. These factors indirectly influence grade factors such as fiber tensile strength, length, and coarseness. Jute, a tropical crop, requires about 30 cm of rainfall per month in its 3 to 5-month growing period. Less than this amount of moisture will reduce fiber length and diameter and thereby lower the quality of processed fiber.

Hemp, a temperate zone fiber crop, has declined in importance as a land use alternative. Cultivars are available which produce fiber of superior quality. Other factors such as temporary water stress, soil deficiencies, and retting problems tend to negate cultivar advances in quality potentials.

Flax is grown for both its structural fibers and its seed oil. Breeding developments have demonstrated gains in yield and quality of one or the other crop product. Cultivars which produce the best quality fiber have few, if any, branches, and are planted in thick stands. Fiber length is maximized and fiber diameter minimized to obtain the best quality of fiber, and this quality is reflected in the growers choice of cultivars, cultural practices, and even areas of growth. Areas having a moist climate with extended periods of cloudy weather produce the best quality of flax fiber.

SUMMARY

Natural fibers derived from crops grown specifically for their fiber production potential have recurring dimensions that reflect genetic, cultural, and processing influences. All are cellulosic fibers, with cotton fibers being almost pure cellulose and structural fibers being mostly cellulose. Fiber dimensions and degree of cellulosic deterioration determine the value and end-use potential of natural fibers.

Field weathering in cotton and improper decortification of structural fibers are the primary causes of cellulosic deterioration among natural fibers. In general, natural fibers that are mature are more resistant to deteriorating agents than are immature fibers.

Length of natural fibers is under genetic control but the gene controlled base length of a fiber may be reduced by any factor that reduces plant activity while the fiber length is being formed. In cotton this period is the 3 weeks following anthesis for individual bolls and the total blooming period of 8 weeks for the whole crop. In structural fibers this period coincides with growth.

Fiber diameter is also under genetic control but is more vulnerable to reductions induced by environmental and cultural agents than is length. An individual boll of cotton is vulnerable for 40 to 60 days, beginning about 21 days after anthesis, and the cotton crop as a whole is vulnerable for up to 120 days. structural fibers are affected similarly. Fiber diameter normally reflects differentials in width of so-called daily growth rings as found in cotton fibers and as found to a lesser extent in structural fibers. Any factor that reduces the photosynthetic activity of the plant during a single day will result in a reduced band of cellulose being deposited inside the fiber.

Cotton fibers are harvested while still attached to seed and generally are harvested with machines which incorporates considerable trash into the seedcotton. Seed are removed from the lint by stationary machines called gins. The gins also attempt to remove the incorporated trash. Improper ginning can cause some fiber breakage and thereby reduce the average fiber length.

Structural fibers are harvested by hand generally and are decorticated with limited help from machines. Fermentation, washing, and drying are involved in decortication. The latter processes require considerable human skills for quality preservation.

SUGGESTED READING

Anon. 1967. Procedings of an Extension Specialists Cotton Quality Conference, Greenville, SC.

Berger, Josef. 1969. The world's major fibre crops, their cultivation and manuring. Centre d'Etude de l'Azote, Zurich, Switzerland.

Brown, H. B., and J. O. Ware. 1958. Cotton. McGraw-Hill Book Company, NY.

Elliot, F. C., Marvin Hoover, and W. K. Porter (eds.). 1968. Advances in production and utilization of quality cotton. Iowa State University Press, Ames.

Lafferty, D. G., and H. A. Alexander. 1971. Effect of cotton variety on producers returns. Arkansas Agric. Exp. Stn. Bull. 764.

Pate, J. B., C. C. Seale, and E. O. Ganstad. 1954. Varietal studies of kenaf, *Hibiscus cannabinus* L., in south Florida. Agron. J. 46:75-77.

Ramey, H. H., Jr., J. H. Turner, and S. Worley, Jr. 1975. The 1972 regional cotton variety tests. USDA Publ. ARS S-62.

Robinson, B. B., and F. L. Johnson. 1953. Abaca, a cordage fiber, USDA Agric. Monograph No. 21.

Part III. Harvest and Storage Effect on Quality

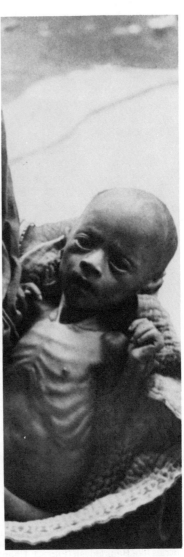

Chapter 8

Quality Preservation During Harvesting, Conditioning, and Storage of Grains and Oilseeds

J. E. FREEMAN

CPC International, Inc.
Englewood Cliffs, New Jersey

Man has been harvesting and storing seeds for future use as food or feed for thousands of years. This practice started in some of the less humid regions of the world. In these areas seeds dry naturally to moisture levels that minimize the risk of spoilage from molds—one of the most serious storage problems in many areas of the world today. Mold damage is a particularly serious problem in tropical areas, whose warm and rainy conditions are favorable to mold growth.

The technology of grain and oilseed harvesting, conditioning, and storage has undergone revolutionary changes in the last hundred years; many of the most important have occurred quite recently. Recognition of the critical role that molds can play in the deterioration of stored grains and oilseeds has had a major influence on these technological advances.

In this chapter agronomic plant seeds that are harvested for food or feed will be designated as "grain or oilseed crops"; the identity "seed crop" will be reserved for seeds intended for planting. This is an important distinction. In general, much more rigorous and costly quality control procedures can be justified with the more valuable seed crops than with comparatively lower priced grain and oilseed crops. Also, for simplicity, the term grain will sometimes be used to represent both cereal grains and oilseeds.

COST OF HARVESTING AND STORAGE LOSSES

Annual harvesting and storage losses are staggering, largely as the result of poor grain management practices. Reliable data on specific losses are scarce and estimates differ considerably. But, no one familiar with the subject questions the seriousness of the problem.

Consider the implications of the following extracts from an article written by an expert in the field, J. L. Ranft.

"The statistics of food waste add up to an appalling spectacle in a world where millions of men, women, and children are continuously hungry. For example, insects such as the granary weevil, Indian meal moth, rice weevil and khapra beetle consumed or spoiled enough grain in 1968 to feed 130 million people ...some 55 million people could be fed for a year on the grain annually consumed in Africa by various kinds of 'freeloaders'—rats, birds, mites, microbes...In central Africa half of the stored sorghum, a staple cereal, is lost to insects every 13 months...In Pakistan grain remaining in storage longer than 10 months is well on its way to losses of 20% and more from insect infestation."

"Rodents destroy upwards of 50% of the food crops in some parts of the world. India, in particular, suffers substantial grain losses from their depredations. One source estimates that India's rats devour or spoil about 10 million tons of grain a year—enough to fill a railroad train almost 3,000 miles long...Damage to rice and other crops by rodents has long been a major agricultural problem in the Philippines. In the 'rice bowls' of Central Luzon and Cotabota, rats consume up to 20% of the expected yield, or more than the annual consumption of a million people...Even the developed nations are not free of crop ravages by rodents: in the USA rats eat around 5.3 million tons of grain annually, and in Australia the common house mouse multiplies at an exponential rate in years of bumper harvests and gorges on the surplus grain stored in open pits."

Losses referred to by this author are principally those resulting from storage pests, such as rodents and insects, that may be classified as consumers. Using this classification scheme, microorganisms are recognized as decomposers. Losses, primarily in quality, caused by decomposers are in many cases just as great as those caused by consumers. But damage by decomposers is much more insidious and consequently even more difficult to measure or estimate.

With the perilous food situation in the world today, no other problem seems deserving of more immediate attention than stored grain losses. The amount of available food is literally a matter of life and death to people of developing countries (Fig. 8.1). To people of the

Fig. 8.1—Hunger is a major problem in Bangladesh, where it is estimated that over half of the 76 million inhabitants do not receive enough food. (Courtesy Agency for International Development, U.S. Department of State).

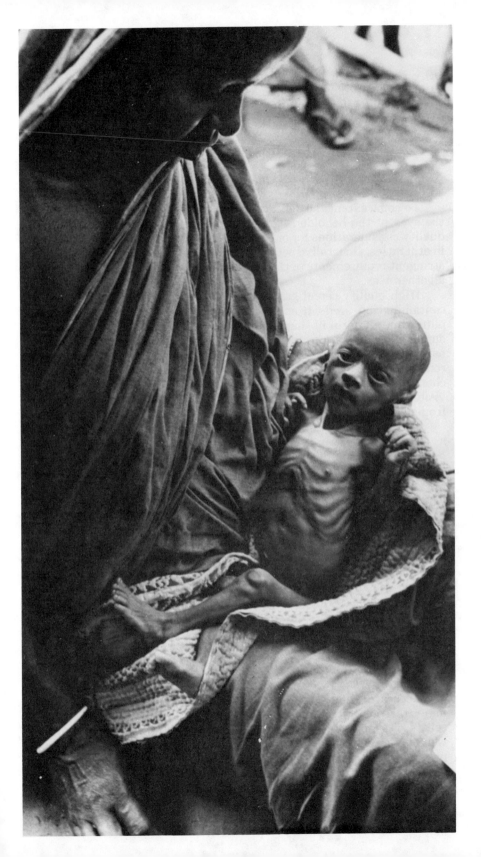

USA and other developed countries the global supply and demand situation has profound economic implications, and quality and wholesomeness of the food supply is a matter of worldwide concern.

QUALITY DEFINITION AND MEASUREMENT

General Quality Determinants

To some extent the definition of quality, like that of beauty, lies in the eyes of the beholder. That is to say, factors that are important quality considerations for one particular end use may not be important in other uses. Generally speaking, however, seeds that satisfy the more stringent requirements of the most demanding end use will be good for all other uses.

High quality cereal grain or oilseeds will 1) be essentially free of contaminating seeds from weeds or other crop plants, 2) be firm, plump and uniform in shape and size, 3) contain low levels of physical damage, 4) have a high percentage of viable seeds, 5) be uncontaminated by insects and mites or their excretory products, 6) be free of rodent feces, urine or hair, 7) be free of toxic pesticide residues or other chemicals, 8) have low levels of microbial contamination with no visible damage from their growth, and 9) be free of toxic metabolites from microorganisms.

The producer of the crop normally judges quality by such factors as color, weight per unit volume, odor, and moisture content. Ordinarily he will be cognizant of the effects of mold or insects only when the degree of infestation is extensive. As the crop moves through market channels, other quality parameters are recognized. The type of measurements that are made are influenced by the needs of the buyer, time and cost required to make the measurements, background of specialists involved, equipment available to analysts, and various other factors.

Extensive quality measurements at any stage in the marketing chain are uncommon because of 1) cost and time involved, 2) inadequacy of some present methods, and 3) inability of users to relate quality measurements accurately and quantitatively to their own specific needs. Simpler and more reliable quality evaluation procedures are definitely needed.

Direct Quality Indices

Moisture

Accurate knowledge of moisture content is important for two major reasons. First, moisture in seeds has no real value. Buyers and sellers of grains and oilseeds therefore must know how much of the weight in any given lot is comprised of water in order to establish a fair price. Secondly, moisture content is closely related to keeping quality of the crop in storage, primarily because of the relationship between moisture content and the growth of storage fungi.

Small measurement errors in the critical moisture range can mean the difference between spoilage or safe storage. Moreover, if moisture content is lower than the measurement indicates, the seller may be unfairly penalized. Conversely, if moisture is higher than the measurement indicates, the buyer may be paying grain price for water.

Electronic moisture meters are commonly used for measuring moisture content of grains and oilseeds. Moisture meters in current use are relatively inexpensive and easy to use. They give quick and reasonably reliable measures of average moisture content of the sample if operated properly and if moisture content is not too high or too low.

Moisture measurement by oven drying is accurate at any moisture content, but methods based on this principle are not rapid and require an analyst. Frequently a sample is not truly representative of the lot being sampled. For maintenance of quality, average moisture values may be meaningless. If some of the seeds in a lot have a moisture content too high for safe storage, they can provide a site of entry for storage molds that can eventually spoil the entire lot.

Unequal moisture distribution can result from any of several factors. Lots of differing moisture content commonly are placed in the same bin simply for convenience. It is not always practical to segregate every lot according to moisture content. Even seeds from different locations in the same field will differ in moisture content. Seeds, especially those of small grain cereals that are exposed to environmental conditions with very little or no protection, may differ widely in moisture content from one time of the day to the other. Moisture content may be higher in the morning after a heavy dew than later in the day. Seeds from different parts of the same plant may differ in moisture content by several percent. Artificial drying can create further variability; unequal drying cannot be avoided. Further moisture shifts can take place in storage.

Even under best conditions some variation in moisture content of individual seeds in a given lot is unavoidable. Variation also results from conscious planning. Although inadvisable from the standpoint of keeping properties, deliberate mixing of lots of differing moisture content to achieve a desired average is common practice—both at the farm and after the crop leaves the farm.

Planned mixing is practiced for two different reasons: 1) to achieve an average moisture content that is believed to be safe for storage and 2) to achieve an average moisture content near the maximum allowed for a particular grade.

Many grain handlers mistakenly believe that when lots of differing moisture content are mixed, moisture content of kernels from each source will equilibrate at a common level corresponding to the average of the mixture. Kernels from the separate lots of differing moisture content approach, but, in practice, do not attain a common moisture content. Consequently, that part of the grain mixture that was initially higher in moisture content may not reach a level safe for storage even though the "average" moisture content is below that at which molds are expected to grow (Fig. 8.2). The difference in final moisture

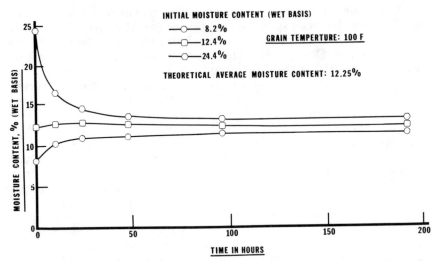

Fig. 8.2—Typical curves showing time response of moisture content for each of the grain fractions in a mixture of high- and low-moisture shelled corn. From White et al. (1972).

content will be greater if the initial spread is large. Persons who have not learned this important point have frequently reported disastrous storage losses that were unexplainable to them.

Grains and oilseeds usually are traded in commerce on the basis of official grade standards that permit some range of moisture content for each grade. It is standard practice, especially by larger elevators that have good blending facilities, to mix lots of differing moisture contents to get averages near the maximum allowed for each grade. Although undesirable from the standpoint of keeping quality, this makes good economic sense. It permits, for example, the following kind of appealing business arithmetic, using soybean [*Glycine max* (L.) Merrill] as an example.

The maximum percentage of moisture allowed in U.S. grades 1, 2, and 3 of soybean are 13.0, 14.0, and 16.0, respectively. Suppose a merchandiser buys two lots of soybean, one of which has 13.2% moisture (U.S. No. 2 range) and the other which has 14.2% moisture (U.S. No. 3 range). Let us assume in this case that, with the exception of moisture content, lot number 2 also meets U.S. No. 2 standards.

Grade is determined on the basis of the grade factor that is most detracting—in this case moisture. Therefore, lot number 2 must be graded as U.S. No. 3 and may be purchased at a discounted price. The buyer of these two lots of beans who has facilities for blending can mix them in proportions to meet U.S. No. 2 standards. This blending, of course, allows reselling the U.S. No. 2 lot without loss of grade while at the same time selling as U.S. No. 2 grade (and at a commensurately higher price) some of the beans originally purchased as U.S. No. 3. As a matter of fact, in this case, 80 parts of the lot that graded U.S. No. 3 could theoretically be blended with only 20 parts of the lot that graded U.S. No. 2 to produce a mixture that will meet U.S. No. 2 standards.

In practice, however, the blender ordinarily would not attempt to blend to the limit but would allow for blending imperfection, sampling and inspection errors, etc. In the illustrative case, for example, a 50:50 blend might be attempted. Even at this level of blending, the financial incentive is significant.

By taking all grade determinants into consideration and practicing judicious blending, elevator operators have a large "standardizing" effect on quality. Some inadvertent standardization is unavoidable from the mixing that results from purely practical grain handling considerations of producers and others. But the planned mixing is different. It not only tends to cause an unnaturally large proportion of the crop that reaches the ultimate consumer to fall into certain grades, but it can also cause a skewed distribution within the grades, with detracting factors tending to be close to the maximum allowable limit.

Although blending has been portrayed as being a generally poor practice, particularly from the standpoint of moisture content, there is a positive side of the issue. Uniformity and predictability of raw material quality is desirable, and blending is sometimes practiced for this purpose. In wheat (*Triticum aestivum* L.), for example, blending of the unprocessed grain and subsequent blending of flour produced by milling commonly is practiced to insure more uniform properties of the flour.

Density

Two different density parameters are important—density of the grain bulk (bulk density) and density of the individual seeds (specific density or true density). Bulk density commonly is used in establishing commercial grades. Bulk density, or test weight as it is known in grain grading terminology, is a function of both true density of the individual seeds and of their packing characteristics. Harvesting, handling, and drying methods can affect both true density and bulk density, the latter of which—by virtue of its relationship to market grade—can have an influence on the price that will be fetched in the marketplace.

Use of the bushel as a measure for grain trading is misleading. A bushel is a volumetric measure. Grain is actually bought and sold by weight, not by volume. A standard weight measure, e.g., 27.2 kg (60 lb) for wheat, is equated with a bushel for trading purposes. If test weight (bulk density) measurement shows that a volumetric bushel would weigh less than the standard bushel weight (low test weight) a discount is applied to penalize the seller for presumed inferior quality. The standard bushel weight differs for different crops. This practice adds to the confusion. The situation would be much simpler if all grains and oilseeds were sold by the hundredweight or by some other common weight measure and the bushel measure could be forgotten.

Test weight measurements are made without any correction for moisture content. As a result, moisture effects are taken into account twice in grade determination since moisture is considered both as an independent grade factor and frequently is the major factor responsible for variation in test weight. This statement is especially true in the starchy cereal grains, which commonly are harvested at quite high moisture contents. Water by definition has a specific gravity of 1.0, whereas starch and protein, the principal constituents of cereal grains, have specific densities of about 1.1 and 1.5, respectively. Consequently, seeds with high moisture content tend to have a lower density (weight per unit volume) than seeds of lower moisture content.

When moisture is removed by drying, kernels shrink to occupy less volume. The extent of volume shrinkage and concomitant increase in true density of the individual seeds can vary for individual lots of a given grain type. Such differences may be a result of cultivar or may reflect differences in methods of harvesting, handling, and drying. Seeds or kernels of lower density after drying may or may not be of inferior quality to kernels or seeds of higher density depending, in part, upon the intended use.

Seeds also can differ in true density independent of moisture content. Kernels that are mature and sound will have high true density, whereas immature seeds and seeds damaged by insects and disease will have a relatively lower true density at a common moisture level.

Physical Damage and Admixtures

Soundness (freedom from breaks, cuts, or bruises) is the most important quality index. A given grade permits a tolerance for different classes of unsound kernels and amounts of foreign material, both individually and collectively.

Some admixture with weed seeds, other crop seeds, chaff, etc., is unavoidable, from a practical standpoint, during the harvesting operation. Any impurity can impair the usefulness of the crop to some extent and, if present in sufficient quantity, may render it unsuitable for some uses. Some weed seeds, e.g., crotalaria (*Crotalaria spectabilis* Roth) and jimsonweed (*Datura stramonium* L.), are toxic and must be removed before the crop is suitable for use as food or feed.

Even some crop seeds are toxic and may present serious contaminant problems. Castorbeans (*Ricinis communis* L.), for example, contain a highly toxic substance that must be removed by appropriate processing. Volunteer castorbean plants frequently appear in cotton (*Gossypium hirsutum* L.) fields when cotton follows castorbeans in rotation sequence in some areas of the southwestern U.S. If the volunteer castorbeans are not rogued, some of the seed will be harvested with the cotton and may eventually contaminate the cottonseed by contributing a toxic substance to the cottonseed oil that is recovered for food use.

Impurities and broken kernels occupy void space in a grain mass, restricting air movement and promoting unequal moisture distribution. These conditions favor microbial activity and resultant quality deterioration.

The damage that is readily apparent in the form of chipped, cracked, abraded, broken, or split seeds is only part of the total damage. A large amount of internal damage is not detected by external observation.

Mechanically harvested seeds of sound appearance may deteriorate much more rapidly than hand-harvested seeds, illustrating the debilitating effects of inconspicuous internal injury. One of the best ways to expose internal damage is to peel the outer pericarp or seedcoat away and stain the seed with 2,3,5-triphenyl-tetrazolium-chloride (tetrazolium). This substance is colorless but is converted to the stable, non-diffusible red formazon by living cells. Therefore, coloration of a cell by tetrazolium is a definite indication of its viability; necrotic cells remain uncolored (Fig. 8.3; see p. 212A).

X-Rays can be used to reveal the presence of internal stress cracks and insect eggs. Scientists have attempted to develop a practical automated method for making quantitative measures of stress cracks in wheat, using x-rays, but to date a method suitable for routine use has not been devised.

Physical damage that occurs during mechanical harvesting predisposes the crop to more rapid (as much as several-fold) quality deterioration in storage.

Heat Damage

Two types of heat damage can occur. One of these has been known, but not well understood, since ancient times. This is the heating that sometimes occurs in stored bulks of seeds or grains. For many years this type of heating was assumed to occur "spontaneously." Today, we know that this heat is generated by biological respiration, most of it being that of microorganisms growing on or in the seeds. When hexose sugars are being utilized by living organisms as a source of energy, heat and CO_2 are produced according to the following equation:

$$C_6H_{12}O_6 + O_2 \rightarrow 6\,CO_2 + 6\,H_2O + 673\mathrm{Kcal}$$

In the early stages of heating by respiration, seeds lose their ability to germinate and begin to darken in color. If nothing is done to stop the heating, the seeds will continue to darken until they become a black mass unsuitable for almost any use (Fig. 8.4). This type of damage can be prevented by maintaining low moisture content during storage.

The second type of heat damage is of recent origin and results from improper artificial drying. Seeds damaged by excessive heat in drying have reduced viability, are darkened in color, and may have blistered pericarps or seed coats. If heat damage is extreme, the seeds may actually explode or partially pop. Dryer damage can adversely affect both true density and bulk density.

Fig. 8.4—Extreme case of heat-damaged soybean (Courtesy of H. H. Kaufmann, Cargill).

The consequences of dryer damage are more serious with some crops than with others. Also, dryer damage may reduce value of a given crop more for some uses than for others. One of the first indications of dryer damage is a reduction in viability. Whereas this reduction is of utmost importance with seed crops intended for planting, it does not have a direct effect on the value of most grains or oilseeds intended for feed or industrial processing. Perhaps the one most significant exception is malting barley (*Hordeum vulgare* L.), for which good germination is essential.

One of the most heat-sensitive crops is wheat. Excessive heat during drying can damage the endosperm protein, impairing the suitability of the flour for breakmaking. Since most wheat is milled to produce flour, heat damage in wheat takes on added significance.

Damage to protein (case hardening) in corn (*Zea mays* L.) diminishes its value for wet milling by rendering the separation of starch and protein more difficult. Excessive heat can even impair utility of a crop for non-ruminant feed by destroying a portion of some of the nutritionally essential amino acids. Heat damage severe enough to cause impairment of feeding value is rare, however, and is manifested visually in the form of darkened color and scorched or blistered kernels.

Indirect effects of dryer damage may be more important than the direct effects. Reduction in viability makes the grain more susceptible to invasion by molds and subsequent deterioration. Brittleness caused by the effects of high drying temperatures leads to more breakage in handling.

Mold Damage

It is virtually impossible, and definitely impractical, to prevent seeds from being contaminated with molds. But it is possible to prevent molds from developing to any significant extent. If mold growth is permitted on seeds, value loss can result from any of several effects: 1) decrease in viability; 2) discoloration of all or part of the seed; 3) loss or modification of seed dry substance; 4) selective destruction of certain seed tissues, e.g., embryo of cereal seeds; 5) development of objectionable odors; 6) production of toxic substances that may be harmful to man or livestock; and 7) on rare occasions, "spontaneous" combustion.

Although mold contamination is not always readily visible, it is common to find mold growth on some portion of the seeds in commercial lots. Official grade classifications recognize such damage as derogatory to the value of the grain. The U.S. Food and Drug Administration has traditionally objected to the use of any moldy products for food because such material is decomposed and therefore adulterated.

Not only is mold contamination frequently undetectable by the naked eye, but commonly it is undetectable even with the aid of an ordinary light microscope during early stages of invasion. In cereal grain, for example, the embryo may be weakened or dead before evidence of the invading mold is detected with a light microscope. Discoloration may not be obvious until the invasion has progressed to the point of decay.

Insect Damage

Insect damage to stored seeds is common, but the full extent of the damage is usually concealed to the layman. Various techniques may be used to expose "hidden" insect damage. As mentioned earlier, x-ray is one such technique. Insect activity also can be detected by an audio technique employing a highly sensitive microphone and amplifier. Infrared analysis of CO_2 respired by insects also can reveal the presence of insects.

Loss of weight and nutrients from insect feeding is only one aspect of stored grain losses from insect activity. Other aspects include various types of contamination (insect fragments, excreta, etc.), dissemination of molds and bacteria throughout the grain mass, and creation of favorable conditions for development of microorganisms caused by moisture and heat produced by insect activity.

Grain lots with gross insect contamination and damage may be placed in a special category, designated by official grading procedures. Insect damage is of particular concern with grain crops, such as wheat, that are used primarily for food. Buyers of grain for ultimate use in food resist contamination by insects to the maximum practical extent.

But since it is impossible to avoid insect contamination entirely, provisions may be made to treat all lots so as to rid them of possible infected kernels. In flour mills, for example, it is common practice to pass the wheat through an impact mill (Entoleter) in which the impact force is sufficient to fracture and pulverize kernels that have been weakened by insect burrowing but insufficient to break sound kernels. The broken, insect-damaged kernels, which may also contain living insect larvae or eggs, can then be separated from the intact, sound kernels.

Chemical Composition

Chemical composition is important for almost all uses but ordinarily is unknown to both buyer and seller. One of the reasons is that analytical methods frequently are too time-consuming and expensive, particularly when both expected compositional differences and relative value differentials are small. In addition, some analytical methods are not sufficiently reliable for identifying small differences.

Where composition is of utmost importance in determining suitability and value for certain uses, compositional data are essential to the establishment of a fair market price. For example, protein content of wheat is important in determining the value of wheat for various uses. Much effort has been devoted to the successful development of quick and simple, but accurate, analytical methods for determining protein content of wheat.

Chemical tests are not all quantitative; qualitative tests may be just as important. Properties of wheat protein and overall baking properties of bread made from wheat, for example, are equally as important as percent protein in the whole grain. Tests for measuring these properties may be quantitative or qualitative. Wet millers process some corn with "special" starch types. These specialty corns are generally grown under contract and their identity preserved through handling and storage at a considerable extra cost to the buyer. It is essential to confirm both the type and purity of such grain before it is accepted for the intended use. Peanut (*Arachis hypogea* L.) is frequently adjudged to be unacceptable for food use because of undesirable flavor characteristics, usually after roasting.

New analytical methods that permit nearly instantaneous measurements of several chemical constituents have been developed. In particular, instruments based on the principle of differential reflectance of near-infrared light of the seed constituents are showing considerable promise. If these methods prove to be sufficiently reliable and economical, they may lead to an entirely new system of grading and pricing of grain crops.

Indirect Quality Indices

Sensory Judgements

Physical appearance is an empirical quality determinant and cannot be measured quantitatively. Nevertheless, it is the most commonly used criterion of quality, and its practical importance cannot be overlooked. One experienced in judging grain crops can tell a great deal from physical appearance. Generally speaking, seeds that score high on all of the direct quality indicators will have good eye appeal. They will be bright, plump, and free of physical defects. For example, corn that is artificially dried, even under best conditions, ordinarily is not as bright and shiny as corn that is dried naturally on the ear.

Physical appearance is related to most of the direct quality indices that have been considered and some of them—e.g., heat damage—are judged almost entirely by physical appearance. Even moldiness, one of the most common and most serious quality defects, is judged on two sensory properties—appearance and odor.

Although it may not have anything to do with quality per se, color is, of course, judged by appearance. From color the experienced judge may be able to recognize the cultivar and deduce other useful information which may be related to quality. Some users have color preferences; official grading standards commonly recognize color classes.

Viability

Seeds with low viability are suspect, but may be suitable for most uses. Viability may be low because of freezing of moist and sometimes immature seeds, mechanical damage during harvest or handling, abusive drying, invasion by microorganisms in the field or in storage, natural aging of seeds or some combination of these and/or any of a number of other factors.

Many buyers routinely check viability of grain they purchase and may use the results to guide them in a selective buying plan. Low germination indicates that something is wrong and is a reasonably good indicator of incipient spoilage. Decreasing germination in storage indicates a high probability that the seed is being gradually invaded by fungi and may soon reach a point where a further increase in mold growth can be sudden, rapid, and disastrous.

Enzymatic and Related Activity

Rapid methods for measuring enzymatic activity may sometimes be more useful than simple germination tests. As discussed previously the topographical tetrazolium test is especially useful for identifying dead or damaged regions of the embryo. This test makes it a good diagnostic tool for determining causes of reduced viability. It is used for this purpose in seed quality laboratories. A seed quality analyst can determine if damage was caused by freezing, crushing, or other factors.

Fat Acidity (FAV)

Efforts have been made to relate the acidity of the fat or oil extracted from seeds to quality deterioration in storage, but this measure appears to be of limited value. Some lots that have deteriorated considerably have low FAV and, conversely, lots of apparent good quality may have a comparatively high FAV.

Fungi appear to be largely responsible for fat acidity. The amount of fat acidity apparently varies with different species of fungi and perhaps even with strains of fungi within a species. And the fungi may produce free fatty acids by breaking down the oil in the seed and then may consume the fatty acids for their own energy source. In this case the free fatty acids would, of course, not be measured. Therefore, the true extent of damage to the oil would be underestimated if this technique was used alone.

Sugar Inversion

Mature, sound seeds contain a small amount of sucrose, part of which may be inverted into glucose and fructose in storage. The proportion of the initial sucrose that is inverted seems to be a reasonably good indication of the extent of quality deterioration and is a measure of quality that has been used for corn with some degree of success.

Grading Procedures

Buyers and sellers must agree on the price of a commodity. To facilitate such an agreement, methods for evaluating quality are necessary. These methods may be formal or informal. The most effective system undoubtedly involves standardized grades established and regulated by government agencies.

Many countries have formal standards for measuring quality and establishing grade designations. Although the specific grades may differ somewhat in different countries they are basically the same. The significant extent of international movement of grains and oilseeds dictates the need for uniformity in grade designations.

Obviously, all producers and handlers of grain should be familiar with official grade standards and procedures. Users of the crops need to go a step further and attempt to relate grade factors to value for the intended use, possibly by relating grade factors to other primary quality determinants such as composition.

The variability of factors that can affect grade—e.g., test weight and amount of damage—is large and continuous. For this reason each grade must allow for some tolerance in levels of each grade factor, and divisions between grades must be chosen arbitrarily. The grading system must not be too complex or it will not be practical. On the other hand, it should be sufficiently complex to take the easily identified quality measures into account and to provide for enough different grades to meaningfully represent normal quality variation among commercial lots.

Soybean cannot be safely stored for any period of time at moisture levels allowed by any of the official U.S. grade designations. This explains their reputation of being difficult to store. The moisture content allowed (13%) with the top soybean grade is about 2% above the level at which they can be stored for more than a short time without loss of quality.

This specific problem may confront the farmer and others in the grain commodities industry. Certain official grade factors seem to have little relationship to value—at least for certain uses. Other important quality determinants are not considered in official grade determinations. Until such time as the official standards are revised, however, buyers and sellers must continue to establish market price on the basis of present standards.

IMPORTANT PROPERTIES OF A GRAIN
OR OILSEED BULK

Physical Properties

Void Volume

The interseed space in a grain mass is designated void volume. The amount of void volume varies with different types of seed. In corn the void volume is about 40%. In other words about 40% of the volume in a container of corn is air and 60% of the volume is occupied by the corn. The smaller, more spherical seed of sorghum (*Sorghum bicolor* L.) pack more tightly and occupy about 63% of the total volume, leaving only about 37% void space.

The amount of void volume also is related to test weight. Seed that is smooth, sound, and dry has less surface friction among contiguous kernels and pack more closely, resulting in higher test weight and lower void volume.

Void volume is important for air movement. If the void volume is occupied by broken grain or foreign material—even in only a part of the stored grain—air movement is impeded and stability is impaired.

Flow

Grain flow is related to quality indirectly and is an important factor to the engineer who designs grain handling and storage equipment.

Physically damaged seeds and seeds of higher moisture content exhibit greater surface friction (higher coefficient of friction) and do not flow as well as sound, dry seed. As a result the slope of a pile (angle of repose) is steeper.

Segregation and Layering

When grain is dropped from a conveyor spout into a bin or silo, the heavy particles (kernels) move readily down the slope to the outside. Irregularly shaped, light particles (pieces) are not as mobile and collect more or less vertically in a column underneath the spout (spoutline). Higher moisture kernels are also less mobile and tend to stay in the spoutline. Weed seeds often are small and irregular in shape and frequently higher in moisture content. Consequently, they too tend to be concentrated in the spoutline. If present in sufficient quantity, the heavier impurities may substantially fill the interseed spaces in a column of grain under the spout. If impurities are tightly packed in the spoutline, air will flow around it when the bin is aerated to reduce temperature. This flow can result in unequal temperature and moisture distributions that, in turn, may create isolated areas favorable for growth of molds, mites, and insects.

Although many impurities collect in the spoutline, those that are light and chaffy have a low terminal velocity and tend to move with the air currents to the walls of the storage bin.

The nature of layering is affected by the physical configuration and dimensions of the bin, as well as by the location of the discharge spout. The amount and type of segregation that takes place has an important bearing on keeping properties of the bulk grain. Even though the total amount of broken kernels and impurities may be small, local concentrations of such impurities can create serious problems. Regions where such materials are concentrated can provide nearly ideal conditions for the initiation of mold growth that can subsequently spread throughout the grain mass.

Temperature

Temperature of a grain bulk is frequently used as an indicator of grain quality in storage—particularly in large, commercial silos that have no provision for aeration. Temperature is usually monitored by thermocouples that are permanently installed in a regular horizontal and vertical pattern throughout the silo. Increases in temperature, or "hot spots," indicate trouble. Positive action must be taken as soon as possible after temperature rises are detected. The best solution usually is to get the grain out of the silo and either dry it or use it immediately.

The main fault with this quality control procedure is that spoilage frequently is quite advanced by the time it is detected. Thermal conductivity of a grain bulk is low, so the heat builds up at the point where it originates. Therefore, spoilage and associated heat that initially develop from insects and/or fungi growing in pockets of the grain mass not near a thermocouple may go undetected during critical stages of initial rapid development.

Because of the poor thermal conductivity of an unaerated grain bulk, heat from external sources penetrates very slowly. Diurnal (day-night) temperature fluctuations rarely affect the temperature of the bulk more than a few centimeters from the surface. At depths of about 200 cm effects of summer and winter temperature cycles are nil. What little effect is noticed is delayed by 2 to 3 months. The atmospheric temperature and the grain and interseed air temperatures are crucial variables for safe and prolonged storage.

Moisture

While average moisture content is needed to assist in establishing a fair price, it has serious limitations in the context of its utility in quality maintenance. If any part of a grain lot is damp enough to support mold growth during storage, then trouble is lurking. Spoilage from unequal moisture distribution in unaerated bins is common and may result in large losses to unsuspecting owners of grain reserves.

In any large mass of stored grain that is not artificially aerated, air currents created by temperature gradients result in moisture transfer. Differences of 2% to 3% in moisture content of samples taken from various points in unaerated bins are common, even though initially of uniform moisture. Thus, areas where moisture content is highest can provide an entry for molds that eventually can spread throughout the grain bin.

Even with the most uniform product, samples from different positions in the same truckload can be expected to differ by as much as 0.5% in moisture content. And these are average moisture values for a relatively large number of individual kernels that comprise the samples. Variation among individual seeds would be expected to be much greater.

Respiration and Useful Life

Both living, dry seeds and dormant mold spores respire, but evidence of their respiration is difficult to obtain. In contrast, actively growing fungi respire vigorously and activity is obvious. Both heat generation and CO_2 evolution can be detected in a grain mass that is supporting active mold development (see respiration equation given earlier). But, respiration rate is not a reliable indicator of incipient spoilage with existing methods.

At any particular moisture and temperature, a given seed lot has a reasonably predictable "maximum useful life" if moisture content is reasonably uniform. This means that a seed lot at a given moisture may become moldy if held longer than a certain number of days. The useful life begins at harvest. If a large part of the useful life is consumed before the final buyer gets the product, it may become moldy under conditions that have been used to safely store newly harvested grain at the same moisture content.

ROLE OF MICROORGANISMS IN QUALITY DETERIORATION

General

Microorganisms that can be found on seed include fungi (molds and yeasts), actinomycetes (unicellular, filamentous organisms), myxomycetes (slime molds), and bacteria. Bacteria, actinomycetes, and myxomycetes are relatively unimportant in quality deterioration in storage. Yeasts require high humidity (minimum of 88%) for growth and are found in abundance only when the grain is held at high moisture. Such grain develops a characteristic fermentation odor. Yeasts are saprophytic (grow on dead plant or animal tissue) and do not harm the grain.

Molds are the real microbial culprits in quality deterioration. Their role was not fully appreciated until recently. Significant damage can occur before the mold is evident. This probably explains why their role as a major factor in grain quality deterioration went unrecognized for so long.

All molds need a source of energy for growth. Typically, they require a source of N, either organic or inorganic, and a carbon source such as carbohydrate or lipid. Seed satisfy these needs quite well. The germ or embryo of cereal grain seed is an especially good substrate as it is a readily available source of minerals, sugars, protein, and oil.

At least 150 different species of fungi have been isolated from seed, but only a few of these are of economic importance. Fungi that grow on seed are divided into two groups—field fungi and storage fungi. Field fungi may be parasitic (grow on living tissue) or saprophytic (generally grow on nonliving organic matter such as the outer tissue of seeds). Fungi that commonly invade seeds in storage are saprophytic.

Field Fungi

The predominant field fungi differ somewhat according to the crop, the region or geographic location, and the weather. The field fungi of major importance, especially in the cereal grains, are species of *Alternaria, Cladosporium, Helminthosporium,* and *Fusarium.*

Field fungi are of relatively minor importance in their effect on grain quality during storage. The most important points to remember are 1) contamination with field fungi tends to be greater in wet regions or seasons than in dry regions or seasons; 2) seed, such as sorghum, wheat, and barley that are exposed to the atmosphere during development usually have a much larger population of field fungi than seeds such as soybean and corn that are protected by pods or husks; and 3) field fungi usually do not grow on stored seed, even if they are heavily infected at harvest, but this depends to a large extent on storage moisture.

Storage Fungi

The storage fungi represent about 12 species of *Aspergillus* (of which only 5 are important), a number of species of *Penicillium*, a single species of *Sporendonema*, and several genera of yeast. These are all common fungi that grow on organic materials and may produce large quantities of spores that serve as a source of inoculum for any inviting material.

Spores are in the air, on harvesting and handling equipment, and especially on the walls of storage bins or silos. The small amount of inoculum present as dormant spores on the outside of the few contaminated kernels, or as dormant mycelium underneath the pericarp or seedcoat of even a few seed is sufficient to initiate spoilage under proper conditions of moisture and temperature. The only practical solution is to maintain conditions that will not permit mold development.

If temperature and moisture are not reduced soon after harvest, it may be too late to prevent quality deterioration. Fungi grow and propagate at exceedingly rapid rates under proper conditions. Cases have been reported of corn being harvested and loaded into a truck in the afternoon and by the next morning being hot from the growth and respiration of fungi. This rapidly consumes part of the crop's "useful life."

Conditions that Favor Development of Fungi

General

The major factors that influence the development of storage fungi on seed are: 1) moisture content of the stored grain and related environmental relative humidity, 2) temperature, 3) oxygen and/or CO_2 levels, 4) the degree to which the grain has already been invaded by fungi before reaching the storage site in question, 5) the amount and type of foreign material present in the grain, 6) the activities of insects and mites, 7) the extent of internal injury and breaks or cuts in the seedcoat, and 8) the length of time the grain is held in storage. All of these factors are interrelated and, to a considerable extent, inseparable.

Table 8.1—Moisture contents for short-term (up to 6 months) storage of grain crops (Courtesy, H. H. Kaufmann).

Crop or Type of Crop	Moisture†
	%
Cereal	13.5
Soybeans	11.5
Flax	8.5
Sunflower	7.5

† Approximate moisture content at which seeds are in equilibrium with 65% relative humidity.

Moisture

Moisture content of the grain is the single most important factor in regulating mold growth in practical grain storage situations. For each of the common storage molds, a minimum moisture content is required for growth. The minimum moisture contents have been determined for most of the common storage fungi.

Actually, the important thing is not moisture content of the seed per se, but rather the relative humidity of the interseed environment. When grain is placed in a closed chamber, the relative humidity of the interseed air space will, in a few hours, stabilize. Equilibrium relative humidity (E.R.H.) is a constant-state relative humidity that is a function of moisture content of the grain, type of grain, temperature, etc. If the grain is not in a closed chamber, but is surrounded by air of a constant humidity—e.g., by aeration—it will gain or lose moisture until E.R.H. is achieved. Curves representing the relationship between equilibrium relative humidity and moisture content of the stored grain are known as absorption and desorption isotherms.

Typical relationships between moisture contents and equilibrium relative humidity of cereal grains and oilseeds are illustrated by Fig. 8.5. The equilibrium moisture content at any relative humidity is greater when seeds lose moisture to reach equilibrium than when they gain moisture to achieve equilibrium. The difference between absorption and desorption curves is known as the hysteresis effect. This phenomenon explains the behavior of blend components as illustrated by Fig. 8.2.

Note that E.R.H. is higher for oilseed than for cereal seed at a given moisture (Table 8.1, Fig. 8.5). This is an important difference between these crops. Oilseed must be dried to lower moistures to assure an E.R.H. at which molds cannot grow.

The most drought-resistant of the storage fungi cannot grow at moisture contents in equilibrium with a relative humidity of approximately 65% or lower. Any seed whose moisture content is below that in equilibrium with a relative humidity of 65% should therefore be safe from invasion by storage fungi, regardless of the other storage conditions.

For storage up to 6 months the best rule, ignoring temperature effects, is to store at a moisture content that will be in equilibrium with 65% to 70% relative humidity. For some common crops this translates roughly to the moisture contents shown in Table 8.1. These relationships are only approximate. Factors such as cultivar and temperature of the grain during artificial drying can cause equilibrium moisture

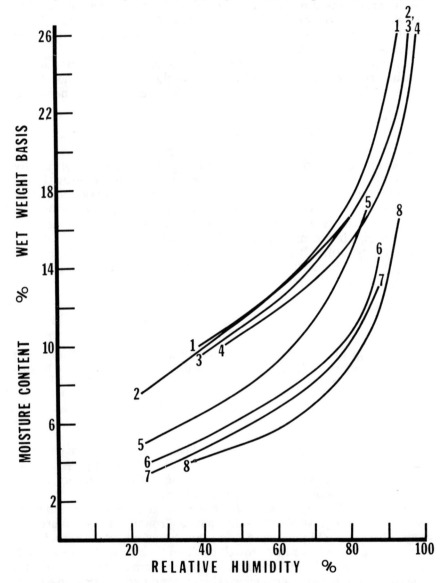

Fig. 8.5—Relationship between moisture content and equilibrium relative humidity of some typical cereal grains and oilseeds. 1. English corn. 2. Cappelle wheat. 3. Barley. 4. American corn. 5. Soybeans. 6. Linseed. 7. Sunflower. 8. Peanuts. From Hunt and Pixton, *In* Christensen (1974).

content to vary by as much as 1%. Beyond 6 months, too many other factors come into play. Even at 13.5% moisture recommended for wheat, "sick wheat" could develop slowly without noticeable heating. ("Sick wheat" is a label commonly given to wheat kernels with reduced germination and darkened germs resulting from mold growth.)

It is generally a good practice to allow some margin of safety in selecting the moisture content for storage. Unequal moisture distribution among individual kernels and throughout the initial grain bulk, possible errors in moisture measurement, and subsequent moisture migration dictate the wisdom of such practice.

Once fungi have started to grow, they continue to grow at a lower moisture content than they otherwise would. Moreover, one of the products of mold respiration is water, which stimulates further mold growth.

Temperature

The storage fungi common on grain and oilseed grow most rapidly at about 30 to 34 C (85 to 90 F). Other factors being equal, growth rate and storage problems decrease as temperature decreases. But temperature effects cannot be considered alone. Interaction with moisture content is paramount, and presence of insects and foreign material are also important. And finally, if mold has already started at a higher temperature, it may continue at a fairly rapid rate even after temperature is reduced. A few hardy species can continue to grow at, or even slightly below, freezing temperature.

Oxygen

All fungi of importance in stored grain spoilage need air. This factor has important implications for grain storage practices to be discussed later.

Other Factors

Mixing of sound, dry seeds with heavily invaded and partially deteriorated moist seeds is a bad practice. This practically assures immediate deterioration and reduction in grade of the entire lot unless the mixture is dried to a low moisture content.

Presence of foreign material can encourage initiation of mold development and foster its subsequent growth. These materials frequently carry a higher spore load than the sound, whole kernels and constitute a more favorable growth medium than does the grain itself. If storage is more than temporary, cleaning to remove broken particles of seeds and impurities such as weed seeds and parts of field insects is highly desirable. Storage or profitable disposition of the cleanings can present problems, however, so blending frequently is chosen as an alternate solution.

Table 8.2—Some recently studied mycotoxins.†

Mycotoxin	Fungi that produce toxin
Citrinin	*Penicillium citrinum* *Penicillium expansum* other *Pennicillium* species *Aspergillus candidus* *Aspergillus niveus* *Aspergillus terreus*
Emetic material of grains	*Fusarium graminearum* (*Gibberella zeae*) other *Fusarium* species
Kojic acid	*Aspergillus flavus* *Aspergillus oryzae* *Aspergillus tamarii* *Aspergillus glaucus*
Ochratoxin	*Aspergillus ochraceus*
Oxalic acid	*Aspergillus niger* *Penicillium oxalicum* Many other species of fungi
Patulin (clavacin, claviformin)	*Pencillium urticae* other *Penicillium* species *Aspergillus clavatus*
Penicillic acid	*Penicillium puberulum* *Penicillium cyclopium* *Penicillium thomii* *Penicillium baarense* *Aspergillus ochraceus*
Rubratoxin	*Penicillium rubrum*
Sterigmatocystin	*Aspergillus versicolor* *Aspergillus nidulans* *Bipolaris* species
Tremorgenic toxin	*Aspergillus flavus* *Penicillium cyclopium* *Penicillium palitans*
Trichothecin	*Trichotecium roseum*

† From Hesseltine (1969).

As discussed earlier, seeds that are mechanically damaged are much more susceptible to deterioration by fungi than are sound seeds. Breaks in the protective pericarp of cereals or seed coat of oilseeds provide easy entry of the storage organism.

Interaction of Fungi with Other Storage Pests

The grain mass is a complex, man-made ecological system in which living organisms and their environment interact with each other. Insects, mites, rodents, and birds can aid and abet the development of storage molds. Feeding and burrowing activity of these pests create sites of easy entry for mold. Their waste products include moisture needed for mold growth; their movements on and in the grain distribute spores.

Mycotoxins

Mycotoxins are toxic substances produced by molds. Much of the knowledge of mycotoxins has been developed only within recent years. Aflatoxin, the mycotoxin that has received most publicity and research attention, was not discovered until after 1960.

Only a few of the molds that can grow on seeds and other organic material produce mycotoxins. Mycotoxins may be secreted into the material on which the molds grow or may be retained within the mold cells and released only with the disruption of the mycelium.

Well over 50 recognized mycotoxins are produced by seed-borne fungi, but only a few of them are of practical concern. Some of the better known mycotoxins are listed in Table 8.2.

Mycotoxins are of somewhat greater importance in areas where less advanced methods of grain conditioning and storage are used and particularly in warm, humid areas. Even in the USA, regional differences are apparent. The incidence of alfatoxin contamination in corn, for example, is much greater in the warm, humid southeast region than in the Midwest. *Fusarium* toxins, on the other hand, tend to predominate in wet, cooler areas.

A discussion of alfatoxin will illustrate important aspects of the mycotoxin problem. Aflatoxin was discovered after thousands of turkey poults and ducklings died in England in 1960 from eating poultry feeds containing moldy peanut (*Arachis hypogea* L.) meal. Intensive study of the problem culminated in the discovery of a toxic substance produced by *Aspergillus flavus,* which was present in peanut meal.

The name assigned to the toxin was derived from the Latin name of the fungus, *(A)spergillus (fla)vus* + (toxin), from which the toxin was isolated. In the few years since its discovery aflatoxin has been found worldwide and in many raw agricultural commodities. Fortunately, however, reported incidence and levels—at least in the developed countries of the temperate zone—have been relatively low.

Aflatoxin has been thoroughly characterized chemically. There are now more than 12 known chemical substances classified as aflatoxins. The most common ones are B_1, B_2, G_1, G_2, B_{2a}, G_{2a}, M_1, and M_2. The first four acount for most of the aflatoxins found on naturally contaminated organic materials. Aflatoxin B_1 is most common and also most toxic. Aflatoxin is both a lethal cell poison, primarily a liver toxin, and a carcinogen (cancer-causing agent) when consumed over a period of time at nonlethal concentrations by some animal species. Aflatoxin also is a mutagen and a teratogen in certain species.

Little definitive information is available on the effects of aflatoxins on man, but there has been strong circumstantial evidence that it has caused human disease. The possibility of human susceptibility has induced the food industry, government, and university laboratories to investigate the incidence and level of aflatoxins in food and feeds, to develop means of preventing contamination, and to explore means of decontamination. Government guidelines and enforcement practices have helped to develop good handling practices that minimize aflatoxin contamination. Voluntary action by industry has also been a major factor in improving grain handling practices and minimizing the risk of injury to man and animals.

The only verified cases of aflatoxin production have been attributed to organisms of the *Aspergillus flavus* group (*A. flavus* and *A. parasiticus*). Organisms in this group are common members of the soil and air microflora and are found in or on various types of organic material throughout the world. *A. parasiticus* is the major species found in tropical and semitropical regions whereas *A. flavus* is found in cooler regions. Although *A. flavus* is classified as a storage mold, it is common in the southeastern U.S., growing on immature corn damaged by insects.

Positive identification of the aflatoxin-producing organisms requires a mycologist, but knowledge of a few important characteristics can assist the layman in identifying samples that are suspected of being contaminated. Members of this category are broadly recognized by their production of greenish-yellow spores that give the fungal colony a conspicuous greenish-gold color (Fig. 8.6 top, see p. 212B). With advancing age the colony color may change to shades of brown. The greenish-yellow color can be changed to a somber yellow-brown by exposing the mold to ammonia vapors. Culture media will influence colony color to some extent.

When examined under a low power microscope, the appearance is as shown in Fig. 8.6 (bottom, see p. 212B). Young sporophores appear white, whereas mature sporophores are covered with greenish-yellow to brown conidia.

Authorities in most developed countries have recognized the danger of aflatoxin and have established and encouraged codes of good practice. In some countries, quantitative tolerances or guidelines based on toxicology and limits of sensitivty of analytical methods have been established.

Regulatory action will be taken with samples at or above the guideline level. Among other things, the guideline takes into consideration the reliability of sampling, analytical methods, and confirmatory toxicity tests. Where guidelines have been established, *blending of contaminated products with uncontaminated products to reduce the mycotoxin concentration below the guideline generally is not permitted.* On the other hand, cleaning of contaminated lots to remove broken kernels and foreign material that may have high levels of aflatoxin is permissable, although this may not substantially reduce aflatoxin content.

One important characteristic of aflatoxin contamination of grains and seeds is that individual kernels or pieces of kernels may be extremely high in aflatoxin content compared with the average for the lot. This makes accurate sampling difficult.

Peanut contamination is of particular concern where peanuts are consumed directly as human food. Invasion by *Aspergillus* occurs after the plants are lifted from the soil and before the nuts are removed and especially in peanuts damaged mechanically or by insects. Quick drying after the vines are lifted greatly reduces infection, and careful inspection before marketing further reduces the possibility of peanuts infected with *A. flavus* getting to the consumer.

Control Procedures

By far the most important single factor in preventing deterioration of stored seeds by molds is low moisture content. Universal recommendations on specific safe moisture levels cannot be made. Decisions must be made locally, taking into account such factors as the type of seed, environmental humidity and temperature, previous history of the seed lot, anticipated storage time, and other relevant factors.

The procedures used in preventing deterioration of stored seed are generally similar for each field crop but may differ to some degree. Using corn as an example, the following procedure is recommended:

1. Harvest when fully mature and operate harvesting equipment in a way to minimize mechanical damage to the grain.

2. Clean thoroughly to remove broken particles and foreign matter such as weed seeds, pieces of stalks and cobs or other contaminants.

3. Aerate to cool and prevent moisture redistribution in the grain mass prior to drying if it is not possible to dry immediately after harvest.

4. Begin drying within 24 hours after harvest and reduce moisture content to 13 to 14% if the grain is to be stored through periods of warm weather.

5. Aerate in storage when practical. Aeration with ambient air may be unsafe if relative humidity is above 70%. This is a particular problem in areas where high humidity prevails over much of the year.

6. Use insecticidal treatments for storage structures. This treatment, of course, is an insect control procedure but may serve indirectly to limit mold growth that might otherwise occur as a secondary effect of insect infestation.

ROLE OF INSECTS, MITES, AND RODENTS IN QUALITY DETERIORATION

Insects and Mites

Importance

Both insects and mites can damage stored seed, but insects are of far greater importance in temperate regions. Some stored grain insects develop inside the kernel while others may bore into wooden storage structures. They are not accessible to contact poisons. Mites tend to be more resistant than insects to pesticides because of the tolerance of the inactive egg and hypopal stage.

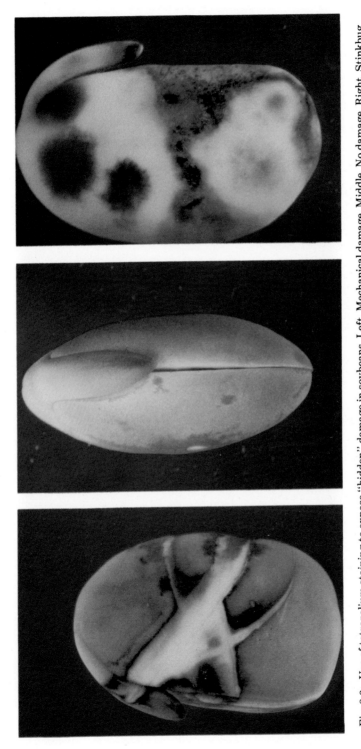

Fig. 8.3—Use of tetrazolium staining to expose "hidden" damage in soybeans. Left, Mechanical damage. Middle, No damage. Right, Stinkbug and disease damage (Courtesy Dr. R. P. Moore, North Carolina State University).

Fig. 8.6—*Aspergillus parasiticus* on corn. Top, Corn kernel heavily infected with *A. parasticius* in the sporulation stage. Magnification, Approx. 6×. Bottom, Detailed appearance of *A. parasiticus* growing on a corn kernel. Magnification, Approx. 200×.

Chemical treatments are directed toward control of insects in the walls of storage structures as well as in the grain bulk itself. Chemical pesticide treatments can be divided into a) fumigation for fast eradication of established infestations and b) contact poisons with residual effects for long-term protection.

Control by Fumigants

Fumigants are useful in controlling insects and mites in stored products. They readily diffuse through the void volume in the grain mass and get into otherwise inaccessible areas and then escape without leaving a residue. Their effect is direct and immediate. However, fumigants afford little residual protection, so it may be necessary to repeat applications.

Some of the chemical fumigant may be absorbed by the grain. The rate of absorption is greater at low temperatures than at high temperatures. On the other hand, the fumigant is more likely to enter into chemical reactions with constituents in the grain at higher temperatures. Some fumigants are more effective at lower temperatures; others are more effective at higher temperatures. This is a complex situation that requires individual attention. Fumigants, as well as contact poisons, generally are more effective and present fewer problems from absorption by the grain at lower moisture levels.

Dosage of fumigants is frequently expressed as a product of fumigant concentration (C) and exposure time (T)—the CT product. Fumigants can be deadly to man as well as to stored grain pests. Some fumigants have strong, characteristic odors. Others have little odor, making them more insidious. To minimize the risk of overexposure to the operator, other strong-smelling gases, or warning agents, are sometimes added to the less odoriferous fumigant. As a general rule, however, detection of fumigants by smell is not reliable.

Containment—i.e., preventing escape of the gas before pests are destroyed—is a major problem with fumigants. A related problem has to do with distribution of the gas within the grain mass. Penetration is aggravated by the presence of impurities that may occupy part of the void volume. Various techniques have been devised to cope with these problems. Gas-proof plastic sheets to cover the storage structure and recirculation of fumigants through the grain mass are most widely used.

For recirculation to be successfully employed, the storage structure must be tight. Conventional aeration systems can be used to distribute the fumigant. To help assure uniform distribution, the fumigant can be dispersed evenly over the surface of the grain or fed into the duct of the aeration system. Air movement should be at least 20 ml/min/liter (0.025 ft^3/min/bu) of grain. For an average size bin, about 5 to 20 min. will be required to displace all of the air with the fumigant-laden air. Recirculation should continue up to about 45 min to assure an effective treatment.

Some of the more common gaseous fumigants are methyl bromide (CH_3Br), hydrocyanic acid (HCN), ethylene oxide (C_2H_4O) and phosphine (PH_3). Methyl bromide is effective against all life cycle stages of stored grain insects and has been widely used. The use of phosphine is increasing. It has excellent penetration ability, does not leave a significant permanent residue, and does not affect the product in any way. Ethylene oxide has been used in recirculation treatment, but it has limitations. Its penetrating power is rather poor, its toxicity to insects is not outstanding, it leaves an undesirable residue, and it reduces germination of many seed. Its use on food grain is prohibited. Hydrogen cyanide is toxic to pests and leaves no appreciable permanent residue, although it is strongly absorbed. It does not penetrate many materials well, however, and has been displaced to a large extent by other fumigants with more favorable properties.

Some chemicals which are liquid at normal environmental temperatures can be vaporized in sufficient concentration to act as effective fumigants in many situations. The principal liquid fumigants used in the grain industry are chloropicrin, ethylene dichloride, ethylene dibromide, carbon tetrachloride, and carbon disulfide. They may be used individually or in combinations of various types and proportions.

Liquid fumigants are particularly well-suited for use on local infestations. The liquid fumigant can be applied by injection through a tube. Enough chemical should be used to provide a lethal dose in the infested area and in the peripheral 1 to 2 m of grain.

Control by Residual Contact Pesticides

Residual contact pesticides have some specific advantages over fumigants. They are generally safer to use, require no specialized equipment for their application, have a higher degree of specificity, and afford some lasting protection. In contrast to fumigants that rapidly dissipate and provide no residual protection, contact pesticides persist on the grain for several months.

Many chemicals are effective as contact poisons against insects, but only a few can be used on grain products. A desirable contact insecticide should possess the following properties: 1) high toxicity to the target pests, 2) low mammalian toxicity, 3) a fairly long residual effect, 4) absence of toxic residues, and 5) absence of undesirable effects on product properties—e.g., production of bad flavor or reduction in viability.

The pyrethrins and malathion (o,o-dimethyl phosphorodithioate of diethyl mercapto = succinate) comprise the principal contact pesticides used on stored grain today. Malathion is widely used on the basis of its broad effectiveness, low mammalian toxicity and relatively low price. Chlorinated hydrocarbons such as DDT [1,1,1-trichloro-2,2-bis(p-chlorophenyl)ethane] and lindane (1,2,3,4,5,6-hexachlorocyclohexane) have been widely used in the past. Restrictions against their use in many countries, however, have caused a decline in their use on stored grain.

Inert mineral dusts have been used since antiquity to combat insects in stored grain. Their effectiveness derives from both a repellance effect and an active effect. The active effect apparently results from a combination of abrasiveness, lipid adsorption, and plugging of breathing tubes.

Addition of abrasive dusts to grain and oilseed presents some problems. In the first place, it creates a potential dust problem when the product is handled. Excessive dust in confined areas can cause explosions. The abrasive dust is also damaging to conveying and handling equipment. And finally, the added dust may result in a grade reduction and hence a reduction in market value.

Rodents

Importance

Rodents are found throughout the world. The most common rodents are rats and mice of various types. The problem with rodents is much less serious in North America than it is in many other areas. And the problem is far less serious today than it was 25 to 50 years ago. The widespread use of anticoagulant rodenticides, better sanitation, and improved building construction have led to a reduced rat population in the USA.

Because of their unusual destructiveness, rats are still a big problem, even in developed countries like the USA. Rats eat about 10% of their weight in food each day and contaminate up to 10 times more with their excreta, thereby rendering it unfit for human consumption. Mice and rats also transmit odors to the grain masses they inhabit. In addition they cause considerable damage to wooden structures, ruin storage bags, and occasionally cause fires by gnawing insulation from electrical wiring.

Droppings and hair from rodents are difficult to remove during cleaning and processing. Consequently, some rodent hair may get into food products derived from grain and oilseed even when they are present at low initial levels. Government regulatory pressure in developed countries, including inspections and seizure of excessively contaminated commodities, has helped to minimize the problem by giving processors and merchandisers special incentive to maintain sanitary conditions on their own premises and to inspect and reject contaminated materials from other suppliers.

Control Techniques

By far the best rodent control technique is prevention through careful handling and improved sanitation. A good rodent-proof storage structure in conjunction with perimeter control to remove rodents from the environment of the building is the best solution. Rodenticides must be used with caution as they can be harmful to both domestic animals and human beings.

HARVESTING, CONDITIONING, AND STORAGE

Harvesting

Mechanized crop harvesting has been an essential development in crop production technology. Yet, mechanized crop harvesting has not come without concomitant problems. With some crops, harvesting technology has temporarily outstripped the technologies of drying and storage. This imbalance has created some special difficulties in quality preservation.

Today, crops frequently are harvested in a different form than previously—e.g., shelled corn vs. ear corn. New handling and storage techniques and facilities are required to accommodate product changes deriving from modern harvesting practices. Some crops are harvested at higher moisture contents and require artificial drying to permit safe storage. Many crops are physically damaged to a much greater extent during mechanized harvesting than they were previously. This damage can reduce the value of the crop for some uses and greatly increase its perishability during subsequent handling and storage. Excessive breakage during harvesting lowers the grade, and hence the price, of the crop.

The amount of physical damage incurred by the crop depends on a number of factors, only part of which are subject to control by the operator. Even when properly adjusted and operated, mechanical harvesters inflict considerable physical damage to the seed, especially when harvesting at high moistures—when seed are both softer and more difficult to thresh or shell than they are at lower moisture content. On the other hand, if seeds are allowed to dry too much in the field they may become more brittle and be more susceptible to breakage or shatter loss during harvesting and handling.

The damage that is readily apparent in the form of chipped, cracked, abraded, broken, or split seed or kernels is only part of the total damage. A large amount of internal damage is not detected by external observation. As mentioned earlier, one of the best ways to expose the internal damage is to peel the pericarp or seedcoat away and stain the seed with tetrazolium (Fig. 8.3). This technique is most useful with oilseed or other dicotyledons.

Despite the quality defects and harvested crop management problems attributable to modern harvesting technology, the proportion of harvestable crops lost in the field or lost in storage as a result of harvesting methodology is far less today than ever before. For centuries, seed crops were harvested by hand with sickles or other hand instruments. Commonly, as in the case of cereal grain, for example, entire plants were harvested and bundled or shocked. Seeds were later threshed or shelled by such methods as hand flailing or by treading under the feet of livestock.

Such harvesting methods are still in use in some parts of the world. But in the more developed countries, mechanical harvesters are used almost exclusively except for rice (*Oryza sativa* L.). In many areas rice is grown in plots too small to justify full mechanization.

The combine is the modern seed-crop harvesting machine. With modern harvesting methods, grain can be harvested quickly and moved immediately to storage for protection from the elements. Weather may not always be conducive to proper field drying and may not permit harvesting at optimum maturity. Also, it is not possible to harvest a crop instantaneously, so the grower has no choice but to harvest some before optimum harvest time and some later.

Recent revolutionary changes in corn harvesting practices in the USA offer a dramatic illustration of the impact that mechanized farming has had on agriculture. In the mid-1950's about one-third of the U.S. corn crop was harvested by hand, and two-thirds by corn pickers that left the grain on the cob. Today, essentially all of the corn is harvested mechanically and earcorn picking has been largely supplanted by field shelling. These rapid developments have drastically altered the entire system of handling and storing and had a great effect on the quality control procedures necessary to prevent deterioration in storage.

When corn is harvested by hand, physical damage to the grain is practically nil. In contrast it is not unusual for 25 to 30% of the kernels picked and shelled at 28% moisture to be visibly damaged. And an even higher percentage of the kernels almost certainly have bruised germs or other cryptic damage. This damage has a substantial influence on quality preservation, as heavily damaged kernels will deteriorate two to five times as fast as sound kernels in storage. Mechanical improvements in harvesting machinery can be expected in response to demands for improved grain or seed quality.

Handling and Conveying

Almost all grain and oilseed are handled several times through pits, belts, conveyors, and other transporting equipment. The seeds are often dropped considerable distances and sometimes hurled at speeds up to 100 km/hr with grain throwers. Screw conveyors are frequently responsible for excessive damage, owing in part to improper operation. To minimize cracking and grinding in screw conveyors, the conveyor should be operated full, using the minimum practical speed. Some reduction in breakage can be achieved by employing grain retarders in downspouts to reduce velocity and minimize discharge impact.

Breakage has become a greater problem with some crops in recent years. An especially large increase in the amount of breakage has been observed with corn and attributed to four major factors: 1) harvest at higher moisture contents, 2) high-impact shelling action in combines,

3) drying with heated air and 4) multiple and high speed handling of corn in market channels. Breakage of corn is more serious at 13% moisture than at 15% moisture and at 10 C than at 21 C. This presents a paradox since shelled corn keeps better at lower moisture and temperature.

Conditioning and Storage

General

Although the principles of crop storage and quality preservation are similar for all grain and oilseed, each crop is different. Storage and handling practices which work well with some crops are unsatisfactory for other crops, both because of different end-use requirements and because of differing inherent characteristics of the seeds.

A major quality problem with cereal grain derives from fissures, or stress cracks, that can form in the endosperm during drying. These stress cracks weaken the kernel structure and result in breakage during subsequent handling. Oilseed are not prone to formation of stress cracks, but splitting of cotyledons is sometimes a problem, particularly with soybean.

The most important distinction between cereal grain and oilseed is the difference in equilibrium moisture contents. Oilseed must be dried to lower moisture to prevent mold growth. Oil does not absorb moisture. Therefore, all of the moisture in an oilseed is concentrated in the non-oil portion of the kernel. So even though moisture content of the whole oilseed must be lower for safe storage, moisture content of the non-oil portion of the oilseed probably is not much different from that of the comparable structure in cereal grain.

Several approaches have been developed to improve storability and maintain quality of seed harvested at high moisture content. The two principal methods employed are artificial drying and aeration. Whereas drying is, of course, used to reduce moisture content to a level safe for storage, aeration serves primarily to maintain low and uniform temperature in order to prevent moisture migration. With sufficiently high aeration rates, however, grain will dry slowly if humidity is not too high.

Grain can be stored safely on the farm for more than a year if a few precautions are taken. As emphasized repeatedly in this chapter, moisture content is of utmost importance. Only clean grain should be placed in storage for extended time, particularly in warm weather. Storage structures should be strong and weather tight.

Aeration

Techniques for forced aeration of stored grain and an understanding of principles of aeration surprisingly are of rather recent origin.

The principal functions of aeration are 1) to reduce the temperature of the grain bulk (molds grow slowly at 4 to 10 C, and mites and insects are dormant) and 2) to maintain a uniform temperature (unequal temperature distribution fosters moisture transfer, which can create conditions favorable for mold).

Aeration can also give a secondary benefit of reducing moisture content. It is possible, for example, to reduce moisture content of corn from 21% to 15% with strong aeration. Moisture is removed from the surface of grain by aeration. So even if the amount removed is small and rate of removal is slow, spoilage may be retarded somewhat by maintaining a lower relative humidity in proximity with surface of seeds. But, aeration usually is not an adequate substitute for drying; particularly if initial moisture content is high. Moist grain can be stored through winter months in cold regions using continuous aeration but it must be dried before warm weather arrives in the spring. Moist grain should never be stored in warm regions, even with aeration.

A less acceptable way of maintaining low and uniform temperature is turning the grain. On the farm this method may actually involve stirring and moving the grain manually with a scoop. More commonly, however, turning is accomplished by moving the grain from bin to bin. Besides being more costly, more troublesome, and less effective, this practice results in increased grain breakage. Moreover, it is much more risky. Incipient spoilage usually is not detected early enough to prevent significant quality deterioration and further spoilage is likely to occur after the grain is moved. Clearly, turning is a poor substitute for aeration, serving primarily to break up "hot spots" and provide minimal aeration.

The void volume, or interstitial space, in a grain mass may be lower if the grain contains small particles—e.g., broken kernels, weed seeds, and dirt. This can impede air movement and is likely to be a particular problem with the "spoutline" where air movement may be essentially nil even with forced aeration. This is one of the principal reasons why cleaning to remove fines is advisable before storing grain. Also, airflow resistance in bulks of some small oilseed—e.g. rape (*Brassica napus* var. *biennis* L.) and flax (*Linum usitatissimum* L.), is much greater than that for cereal grains.

As discussed earlier grain at moistures normally encountered in storage exhibit poor thermal conductivity (low specific heat and high insulating properties). The practical consequence is that if temperature differences exist between lots of grain when they are loaded into the bin, or if grain near the surface or walls of the bin is influenced by environmental temperatures, temperature differentials will persist in an unaerated grain mass. This condition causes air currents to be established. Warm air rises and cool air flows downward. When the warmer air comes into contact with cooler grain, moisture is transferred from the air to the grain. The opposite transfer occurs when cool air comes into contact with warmer grain. Moisture transfer is likely to be greater when moisture content of the grain is higher and/

or when temperature is higher. "Top-sweating" is a common problem caused by warm air rising in the grain and raising the moisture content of the cooler grain it encounters on the top surface of the grain bulk. Under extreme conditions of "top-sweating," moisture may actually condense on the top of the bin and drop onto the surface of the grain bulk.

Aeration does not lower the temperature of all of the grain in a bin uniformly. Rather, a "cold front" is formed. As air is forced through the grain mass, the front slowly moves in the direction of the air movement. All grain behind the front will have the temperature of the entering air. Grain ahead of the front will retain its initial temperature until the front reaches it.

Artificial Drying

For various reasons, natural drying of seed in the field may be impractical. For example, leaving grain in the field too long results in field losses from lodging and shattering. Also, bad weather may prevent harvest. Therefore, artificial drying becomes a preferred or essential alternative.

In practice, the common aim is to remove the minimum amount of moisture necessary to meet a target grade, and to do this at the lowest cost. A preferred objective is to effect at least cost the minimal moisture removal consistent with requirements for safe storage and to do this with minimal damage to the grain.

Seeds are hygroscopic. Consequently, dried seeds can regain moisture by any of a number of methods—e.g., coming into contact with moist air or from moisture shifts that take place as a result of convection currents of the type described in the previous section.

Air temperature is not the critical factor in drying; grain temperature is what counts. The air temperature considered safe for heated air drying is related to the method of drying. In many dryers air flows through a stationary layer of grain causing grain closest to the incoming air to dry first, whereas that near the outgoing air dries later. Moisture content and grain temperature both vary greatly in this stationary layer. This situation is aggravated by using thicker layers and lower rates of air flow.

In other types of dryers, the grain moves over a series of hot and cold air ducts so that the same kernels are not exposed to the hottest air for the entire drying period. In such dryers the grain temperature will be considerably cooler than the temperature of the entering air. Some types of dryers can operate at high air temperatures without significant damage to the grain because the cooling effect of water evaporation keeps grain temperature low.

A technique involving a combination of heated-air drying and aeration is known as "dryeration". Dryeration has been used with rice, corn, soybean, and perhaps with other crops. Grain is moved from the dryer, while still hot, at a moisture about 2% above the desired level. It is then placed in a bin where it is allowed to temper from 4 to 8 hours before it is cooled slowly by ventilation with natural air. This method of drying is reputed to offer some fuel economy as well as to minimize stress crack formation that can result from drying and cooling too rapidly through a critical moisture range. In corn, for example, most stress cracks develop while drying through the moisture range of 19 to 14% and are more numerous when starting from high moistures.

When high moisture corn is dried faster than about 8 to 10 percentage points/hour, the kernels puff or expand. A cavity is formed in many of them and bulk density is reduced. Most grain crops ordinarily are not dried from moistures high enough to cause this problem. Corn is a notable exception.

As more and more water is removed, reducing moisture further becomes more difficult and more expensive. Drying below the maximum safe level for storage is costly in terms of 1) fuel cost, 2) loss of weight in the form of evaporated water that could be sold at grain price, and 3) possible quality loss from scorching and discoloration or from subsequent breakage of the more brittle seeds. Brittleness is a particularly serious problem with artificially dried rice; broken rice generally has a much lower market value than does whole rice.

Other techniques with potential for improving grain drying include stirring, multiple drying zones, and "low-temperature" drying. Low tempeature drying (-1 to 10 C), in particular, has considerable appeal. Grain can be slowly dried over a period of several months by raising the temperature a few degrees above the ambient temperature. One technique that has been used is to capture the heat from the electric motor used to drive a fan for aeration and use it to dry. At a time when gas is expensive and sometimes difficult or impossible to obtain, this technique is especially attractive. But a degree of peril is inherent in low temperature drying that cannot be overlooked. If relatively warm, moist conditions prevail during and after harvest, the drying process may be too slow to prevent spoilage from mold and possible contamination from mycotoxins.

The problem is that a drying "front" develops. Grain near the incoming air, usually at the bottom of a bin, soon dries to reach equilibrium with the relative humidity of the incoming air. The drying air gains moisture until it is saturated. It then moves through the rest of the grain mass with no drying effect. Consequently, grain near the top of the bin, although cooled by the air movement, will be no drier than it was when placed in the bin until the drying front reaches it. This process may require as much as 2 or 3 months. For this reason, low temperature drying should not be used in many areas. Ordinarily, bins are filled partially, the grain is dried, and more grain added. Air flow rates, moisture content, and grain depth all are involved in determining safety of this practice.

Some crops are more difficult to dry than others, and even cultivars of a single crop differ in drying properties. Soybean have always been considered difficult to dry and frequently are not dried. Drying of peanuts presents a special problem in that flavor of the dried product is of major importance. Drying air temperatures of approximately 35 C commonly are used.

Storage

In less developed countries, grain is stored in small buildings, in woven bags and baskets, or in piles out-of-doors. Even in developed countries such as the USA, grain may be temporarily stored in piles on the ground because of a shortage of railroad cars needed to move the crop to market and to the rapid harvesting made possible by combines.

In commercial grain areas, however, grain ordinarily is stored in concrete silos and tight wooden or metal bins. Aeration is recommended for these structures and in situ drying may sometimes be practical.

Hermetic (airtight) storage has not gained wide acceptance, primarily because of the difficulty of producing a completely airtight structure and the objection to the fermented odor of the product for other than feed use. Storage at high moisture has a beneficial effect on feed value for ruminants. Feed efficiency of grain stored at 25 to 30% moisture is increased by 5 to 10%. Frequently, the grain is cracked to improve packing and exclusion of air.

Butyl-rubber silos have been tested successfully in Nigeria, Kenya, and the United Kingdom. Underground pits are most common and have been in use to a limited extend for many years. The rate of heat transfer into or out of grain stored in underground bins is slow (similar to that in large bins and insulated bins) owing to good thermal insulation of soil.

Bins made of concrete with an aboveground dome over a conical pit base are being used to store grain in Cyprus, Kenya, and other locations. Vapor proofed concrete is necessary in areas with high water tables.

Underwater storage has been suggested in recent years and may have some potential for storage of reserves in certain countries—e.g., Japan—where space is limited and water is accessible.

Successful airtight storage depends on the depletion of oxygen in the sealed structure. Oxygen depletion results primarily from the preliminary respiration of molds and insects. As oxygen is depleted, all of the living organisms cease to respire and grow. Oxygen at a level of 2% by volume is sufficiently low to inhibit most insects. At 0.2%, mold activity is halted. Only a few microbes, notably yeast, grow at between 0.2 and 1.0% oxygen.

Both damp and dry grain can be stored in airtight structures. Whereas the properties of dry grain changes little in airtight storage, damp grain (over 16% moisture for cereal grains) is modified to the point that commercial value and suitability for certain uses are affected.

Above about 16% moisture, further anaerobic respiration takes place after oxygen has been depleted, until CO_2 may constitute 95% of the air. In some structures, such as conventional tower or bunker silos that are not completely airtight, some CO_2 escapes to the air. Consequently, the level of CO_2 tends to stabilize at 15% to 25%. Seed stored under these conditions lose viability.

Chemical Preservation

Use of chemical preservatives is restricted to grain intended for feed, being used primarily on grain to be fed by the producer. Special applicators are needed to apply the chemicals.

Most of the commercial grain preservatives are organic acids, used alone or in combination. Propionic acid and acetic acid seem to be of primary interest but formic, sorbic, and others have utility. Low levels of chemical preservatives may be useful in suppressing or inhibiting mold development during low-temperature drying.

One limitation of present preservatives is their corrosiveness. Conveyors, metal grain bins, transportation equipment, and other metal equipment can be seriously damaged. However, grain treated with acid preservatives can be stored in structures that are otherwise unsuitable for grain storage. Because of these characteristics, chemical preservatives are most attractive in situations where grain dryers and conventional storage structures are unavailable or inadequate. Also, there is some incentive to use chemical preservatives with corn to be fed to ruminants. These animals can metabolize the organic acid preservatives as an energy source and they also benefit from the advantage of improved feed value of high moisture grain.

Refrigeration

Since about 1965, significant efforts have been made in the USA to develop refrigerated grain storage technology. But to date, this concept has not demonstrated sufficient attractiveness to achieve commercial acceptance. Cost appears to equal or exceed drying costs, and moisture content is not reduced. Also, refrigeration does not eliminate all risk of spoilage. Mold development on top of the grain mass, resulting from top sweating, is a particular problem.

One potential use of refrigeration is temporary storage. Chilling to 4 C or below immediately after harvest will greatly increase allowable storage time before drying. The cool air will also remove some moisture.

SUMMARY

Poor practices in harvesting, conditioning, and storage of grains and oilseeds are responsible for quantitative and qualitative losses of enormous magnitude. More tragic than the economic loss is the deprivation of a large amount of desperately needed food from millions of hungry and malnourished men, women, and children around the world.

Good management practices are necessary to preserve the inherent quality of a crop as it stands ready for harvest. Harvesting equipment must be adjusted to minimize both field losses and mechanical damage to the seed. Mechanical damage during harvesting can reduce market grade upon which price is based. It also makes the grain more susceptible to further breakage during handling and to spoilage in storage, especially as a result of mold growth.

Grain crops commonly are harvested at moisture contents too high for safe storage. Moist grain can be held safely for short periods in cold weather by cooling with aeration. For prolonged storage, a better method is needed. In developed countries, artificial drying with heated air is by far the most widely employed practice for conditioning moist-harvested grain for storage. As a rule, grain should be dried to a moisture content that is in equilibrium with a relative humidity of about 65 to 70%. This translates to approximately 12 to 14% moisture for most cereal grain and 8 to 10% moisture for most oilseed. Under these conditions, common storage molds do not develop.

Mold is one of the major grain storage problems in many parts of the world today. Molds use part of the grain dry material for their growth and undesirably alter appearance, odor, and composition to reduce overall quality of an affected grain lot. Some molds also produce toxic metabolites known as mycotoxins. If present in sufficient quantity, mycotoxins can be harmful or fatal to animals that consume the contaminated product.

Low grain moisture also retards development of insects and mites. If insects or mites inhabit a stored grain bulk, fumigation or treatment with a contact poison may be necessary for eradication.

Consumption and contamination by rodents is a major storage problem in some areas. The best protection from rodents is a good, tight storage structure. Perimeter control with approved rodenticides may also be necessary.

Although aeration and artificial drying are the methods most widely used in developed countries for insuring quality maintenance in storage, other practices are sometimes employed. Chemical preservation is a new technique that is gaining some acceptance as an alternative or supplement to drying. At low levels, chemical preservatives show some promise for use in suppressing mold development while grain is slowly dried at temperatures, just above ambient temperature.

Moist grain can be preserved by airtight (hermetic) storage or ensiling. Ensiling alters properties, particularly flavor and odor, in ways that render the product unacceptable for some uses. Ensiled grain is used primarily as a cattle feed.

SUGGESTED READING

Anon. 1968. Barley: origin, botany, culture, winter-hardiness, genetics, utilization, pests. USDA Agric. Handb. 338.

————. 1973. Rice in the U.S.: varieties and production. USDA Agric. Handb. 289.

————. 1973. Peanuts—culture and uses; a symposium. Am. Peanut Res. and Educ. Assoc. Inc., Stillwater, OK.

————. 1975. Grains and oilseeds. Handling, marketing, processing. Canadian Int. Grains Inst., Winnipeg, Manitoba.

Arant, F. S. (ed.). 1951. The peanut: the unpredictable legume. Natl Fert. Assoc., Washington, DC.

Bailey, A. E. 1948. Cottonseed and cottonseed products; their chemistry and chemical technology. Interscience, NY.

Brooker, D. B., F. W. Bakker-Arkema, and C. W. Hall. 1974. Drying cereal grains. AVI Publ. Co., Westport, CT.

Caldwell, B. E., R. W. Howell, R. W. Judd, and H. W. Johnson (eds.). 1973. Soybeans: improvement, production and uses. Am. Soc. Agron., Madison, WI.

Christensen, C. M. (ed.). 1974. Storage of cereal grains and their products. 2nd Edition. Am. Assoc. Cereal Chem., St. Paul, MN.

————, and H. H. Kaufmann. 1969. grain storage; the role of fungi in quality loss. Univ. of Minnesota Press, Minneapolis.

Cotton, R. T. 1963. Pests of stored grain and grain products. Burgess Publ. Co., Minneapolis, MN.

Delorit, R. J., L. J. Greub, and H. L. Ahlgren. 1974. Crop production. Prentice-Hall, Englewood Cliffs, NJ.

Doggett, H. 1970. Sorghum. Longmans, NY.

Hesseltine, C. W. 1969. Mycotoxins. Mycopathol. Mycol. Applic. 39:371–383.

Houston, D. F. (ed.). 1972. Rice: Chemistry and technology. Am. Assoc. Cereal Chem., St. Paul, MN.

Hughes, H. D., and D. S. Metcalfe. 1972. Crop production. 3rd ed. McMillan, Riverside, NJ.

Inglett, G. E. (ed.). 1970. Corn: culture, processing, products. AVI Publ. Co., Westport, CT.

————. 1974. Wheat: production and utilization. AVI Publ. Co., Westport, CT.

Jugenheimer, R. W. 1976. Corn: improvement, seed production, and uses. Wiley, NY.

Leonard, W. H., and J. H. Martin. 1963. Cereal crops. Mcmillan, Riverside, NJ.

Markely, K. S. 1950-51. Soybeans and soybean products. Vols. I and II. Interscience, NY.

Martin, J. H., W. H. Leonard, and D. L. Stamp. 1976. Principles of field crop production. 3rd ed. McMillan, Riverside, NJ.

Myasnikova, A. V., Yu. S. Rall, L. A. Trisvyatskii, and I. S. Shatilov. 1969. Handbook of food products. Grain and its products. Israel Program for Sci. Transl., Jerusalem.

Pomeranz, Y. (ed.). 1971. Wheat: chemistry & technology. 2nd ed. Am. Assoc. of Cereal Chem., St. Paul, MN.

————. 1975. Advances in cereal science and technology. Am. Assoc. Cereal Chem., St. Paul, MN.

Quisenberry, K. S., and L. P. Reitz (ed.). 1967. Wheat and wheat improvement. Am. Soc. Agron., Madison, WI.

Ranft, J. L. 1971. Waste—hunger's ally. War on hunger. U.S. Department of State, Agency for International Development, Washington, DC.

Sinha, R. N., and W. E. Muir. 1973. Grain storage; part of a system. AVI Publ. Co., Westport, CT.

Smith, A. K., and S. J. Circle (eds.). 1972. Soybeans: chemistry and technology. Vol. 1. Proteins. AVI Publ. Co., Westport, CT.

Sprague, G. W. (ed.). 1976. Corn and corn improvement. 2nd ed. Am. Soc. Agron., Madison, WI.

Tsen, C. C. (ed.). 1974. Triticale: first man-made cereal. Am. Assoc. of Cereal Chem., St. Paul, MN.

Wall, J. S., and W. M. Ross (eds.). 1970. Sorghum production and utilization. AVI Publ. Co., Westport, CT.

White, G. M., J. J. Ross, and J. D. Klaiber. 1972. Moisture equilibrium in mixing of shelled corn. Trans. ASAE 15:508–514.

Wolf, W. J. 1975. Chemistry and technology of soybeans. Am. Assoc. Cereal Chem., St. Paul, MN.

Woodroof, J. G. 1966. Peanuts: production, processing, products. AVI Publ. Co., Westport, CT.

Chapter 9

Quality of Forage as Affected by Post-Harvest Storage and Processing

L. E. MOSER

University of Nebraska
Lincoln, Nebraska

Maintaining forage quality during harvest, processing, and storage can increase livestock performance and overall feed efficiency. In the past, more emphasis has been placed on reducing grain harvesting and storage losses than on reducing forage losses. Many of the losses and subsequent changes in feed values in forages are subtle. A little leaf loss, soluble carbohydrate loss, or protein loss is hard to detect on the farm. However, such losses lower forage quality much more than the dry matter loss would indicate.

Today, many producers are aware of the importance of quality forage in the production of meat, milk, and wool. Often they are proficient in producing a high yielding, high quality forage crop prior to harvest. Considering the high costs of feed production, producers are at a place where more emphasis should be placed on forage quality changes after harvest. Quality can be quite varied when considering all of the possible storage forms of forages and the many species of plants that are used. Changes in feed value take place from the moment a forage is harvested until it is completely ingested by the animal.

CHEMISTRY OF DRYING

Many forages are dried before they are processed and stored. With the exception of vitamin D, the nutritive value of stored forages is almost always lower than it is prior to harvest. Nutrient losses start immediately after cutting, and some biochemical losses are unavoidable in commercial production. The removal of water as quickly as possible results in the lowest losses in quality. Biochemical losses are often not recognized by the producer even though he recognizes leaf shattering, leaching, or molding losses.

Nutrient Changes

Carbohydrate Loss

Plant cells remain alive immediately after cutting so plant processes and enzymes remain active. Respiration of living plant cells causes a loss of starch, fructosans, sugars, and other readily available carbohydrates and, to a lesser degree, organic acids. Since carbohydrates used in plant respiration are essentially 100% digestible by animals, this loss is important. When the plant dries quickly after harvest, respiration losses are kept to a minimum. As the moisture content of the plant reaches 35%, all cell respiration activity ceases (Greenhill, 1959). If plant cell life is prolonged by wet conditions the respiratory loss is much greater than when plants are dried quickly. A small amount of photosynthesis takes place in cut plants, but the magnitude is insignificant since stomata close with the onset of water stress. As a result, additional CO_2 entry into the cut forage plant is quite low. Furthermore, much of the forage is covered, particularly when windrowed, and receives little sunlight.

With natural drying, forage cut in the evening generally undergoes greater metabolic losses than forage cut during the day. Losses are greater during warm nights than during cool nights, due to increased respiration. Since starch and other carbohydrate storage compounds are hydrolyzed and produce further substrate for respiration, the concentration and quantity of sucrose increases in alfalfa (*Medicago sativa* L.) during the night while starch decreases (Knapp et al., 1973, and Fig. 9.1). Respiration losses of carbohydrates in air-dried forage were around 3.5% during a 24-hour period in a study by Hesse and Kennedy (1956). However, alfalfa cut in the evening had overnight dry matter losses ranging from 7.2 to 11% (Knapp et al., 1973). Holt has suggested that post-harvest metabolic losses in hay are large, and major effort is needed to reduce this source of forage loss. Possibly the oxidase enzymes that catalyze oxidative reactions are active in drying plants since these reactions have been demonstrated in drought-stressed plants. Under stress there is an increase in oxidative reactions. After plant enzymes have lost their activity, other non-enzymatic reactions or other organisms can cause nutrient losses. High initial moisture, high soluble sugar content, high relative humidity, and high temperatures result in the greatest carbohydrate loss. Since structural carbohydrates in the cell wall, such as cellulose and hemicellulose, are undiminished in the drying process, the proportion of the structural fraction during drying is higher because soluble nutrients decrease.

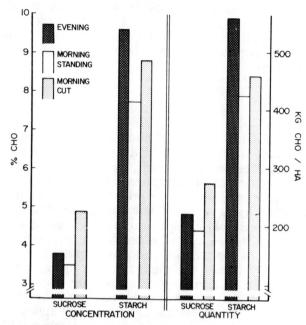

Fig. 9.1—Overnight changes in the carbohydrates (CHO) sucrose and starch in cut and standing alfalfa in May. Note that some starch is hydrolyzed to sucrose in cut alfalfa overnight. From Knapp et al. (1973).

Protein Loss

Protein content is affected less than carbohydrate content by plant metabolic processes during drying. Even if protein hydrolysis occurs, amino acids generally would not be lost. Loss from volatility of nitrogenous compounds is minor under normal drying conditions. However, when forages heat or mold, N losses can occur. Even then, dry matter losses are often higher than N losses in heated or molded forages. This loss can result in an increased amount of N, when measured on a percentage basis, in the damaged forage.

Vitamin Changes

Vitamin A or its precursor, carotene, is affected greatly by the drying process, especially if the drying period is prolonged. Slow drying with warm temperatures enhances activity of lipoxidase, which is the enzyme that breaks down carotene. Photochemical breakdown of carotene occurs under bright sunlight. The yellow-colored carotene, although it normally is masked by green, correlates closely to the green color of chlorophyll at the time of storage. Green forage is high in carotene, while bleached forage is low in carotene. However, even under the best hay-making conditions, carotene loss may be 75% compared with the standing crop. Vitamin E is decreased by the drying process, but the B vitamins are generally not lost to any great extent. However, B vitamins may be readily leached by rain since they are water-soluble. Vitamin D is formed during the drying process by light radiation of its sterol precursor, ergosterol. Vitamin D is at its highest in fresh, sun-cured hay that is not weathered.

Other Losses

Losses of minerals, such as P and Ca, in the drying process are often quite small. However, prolonged field drying exposes forage to the weathering process. As a result, leaching, leaf shattering, and other indirect physical causes of mineral loss may occur.

Some toxic compounds present prior to harvest are not potential hazards after the drying process. Plants such as sudangrass [*Sorghum sudanense* (Piper) Stapf.] containing cyanide-releasing glucosides in the fresh herbage are not toxic after the drying process has been completed. Plant cells become disorganized upon drying, releasing glucosidase, which breaks the sugar from the cyanogenic glucoside. Eventually volatile HCN readily dissipates. Dry hay does not cause the foamy type of bloat because soluble proteins, which appear to be largely responsible, are denatured and lose their ability to stabilize foam. Estrogenic substances, such as coumesterol in alfalfa, which interfere with the estrus cycle in animals and disturb conception, decrease with drying. Sweetclover [*Melilotus alba* Desr. or *M. officinalis* (L.)] can become toxic if molding takes place. Toxicity occurs because coumarin in the plant changes to dicoumarol, a vitamin K antagonist, which reduces the ability of blood to clot.

Speed of Drying

More nutrients can be preserved with rapid, artificial drying than with relatively slow, field drying. Drying forage at high temperature, but not so high that it kills cells, will initially cause more carbohydrate loss from increased respiration. With a more rapid water loss, enzymatic activity will stop quickly. At temperatures of 70 to 100 C, enzymes are quickly denatured and therefore respiratory losses are kept to a minimum. The heat of evaporation keeps forage cool initially, but as drying takes place the forage temperature elevates. If forage is allowed to dry completely at a high temperature (70 to 100 C) a small amount of sugars and proteins are lost, and subsequent digestibility of the forage may be lowered.

Lyophilization, or freeze drying, quickly stops metabolic activity. However, some enzymes are not denatured; if normal temperatures return, along with moisture, some metabolic activity might occur. Under laboratory conditions Smith (1969) reported that forages dried at 100 C for 90 min followed by complete drying at 70 C most closely corresponded chemically to freeze-dried samples. Generally with rapid, artificial drying the levels of carbohydrates, true protein, and carotene, are higher, whereas the crude fiber level is lower compared with field-dried forages. These changes are related to the speed of enzyme deactivation as well as to the increased field losses incurred when dry forages are handled. Weather damage is often avoided when forages are brought in at high moisture and artificially dried.

HARVESTING LOSSES

Various types of forage harvesting losses may occur in addition to the nearly unavoidable losses associated with the biochemistry of drying. Following is a general list of types of harvesting losses, similar to those discussed by Hundtoft (1965a).
1) Cutting losses caused by leaving high stubble.
2) Cutting losses caused by leaves being cut or stripped off by machinery.
3) Field respiration and fermentation when cell life is prolonged or weather conditions cause microbial activity in forage.
4) Leaching and bleaching causing loss of soluble nutrients and vitamins.
5) Leaf shattering losses associated with handling excessively dry forage.
6) Losses incurred during post-harvest storage which can be especially high if heating or molding occurs.

Most losses on the above list are theoretically avoidable. However, unexpected rain causes some of the losses to be unavoidable to the producer.

Weathering Losses

Rain in most regions if often blamed as being the main "problem" in haymaking, Rain can affect forage drying in a number of ways. Rain may 1) prolong the life of plant cells, continuing respiration, 2) leach soluble nutrients from dried plant cells, 3) indirectly cause considerable leaf loss and, 4) cause an environment favorable for microorganisms that cause fermentative losses. Excessive weathering reduces hay quality markedly. Weathered outside portions of large hay packages averaged 33.3% total digestible nutrients (TDN) while the unweathered cores averaged 56.7% (Smith et al., 1974).

Leaching of nutrients is important, but not all losses from rain are due to leaching. Leaching losses occur only from dead plant cells when the protoplasm and membranes have lost their selectivity and differential permeability. Leaching losses increase with the number of rain showers, amount of rain, and with increasing dryness of the hay before it rains. Prolonged exposure of cut forage to the weather will also cause nearly a 100% loss of carotene. Dry matter losses due to leaching

Fig. 9.2—Under semi-arid conditions nutrient retention was better where forage was round baled and left outside compared to bunched or standing forage. Photo by John K. Ward and Steven S. Waller.

per se have varied from essentially nothing to 20% in leaching experiments where rainfall was simulated. As with respiration losses, leaching losses are composed of soluble nutrients that are nearly 100% digestible. If soluble nutrients are leached out, leaving the less digestible plant parts (cell walls or structural parts), overall digestibility drops markedly. Soluble mineral elements may be leached as well, and especially high amounts of K are removed from rain-damaged forage. Feed value of forage left standing in the field may decrease sharply with freezing weather due to the rupture of cells and the leaching of soluble materials with rain. In semi-arid to arid rangelands, standing forages have better winter quality than similar forage left standing in a humid area because of reduced nutrient leaching and forage deterioration.

Since leaves are the most valuable part of forage, nutrient loss is greatly accented by leaf shattering. A loss of dry matter due to leaf shatter is magnified greatly when the reduced feeding value of the subsequent forage is considered. Rain not only leaches soluble nutrients, but also causes extra handling to facilitate drying which, in turn, increases leaf loss. Losses due to microbial activity are another indirect effect of rain. When forages are rehydrated, microbial populations, primarily fungi, are increased. This increase may cause additional carbohydrate loss from fungi respiration.

Generally most of the microbial loss occurs after harvest when hay is stored with excessive moisture; however, under wet conditions a

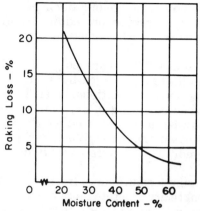

Fig. 9.3—Shattering losses in alfalfa as influenced by moisture content when raked. From Hundtoft (1965b).

considerable amount of microbial and leaching losses may occur in the field prior to storage. Research conducted under semi-arid conditions showed that nutrient retention was much better where forage was round-baled and stored outside (Fig. 9.2) compared to bunched forage or standing forage (Burzlaff and Clanton, 1971). With the round-baled forage, leaching occurred only on the outside of the bale.

Mechanical Losses at Harvest

Leaf shattering, especially with legume hays, can be one of the most serious losses at harvest time. The nutritive value of leaves was illustrated by Hundtoft (1965c) where, in New York, first-cut alfalfa averaged 58% leaves and 42% stems by weight. The stems contained 44.6% crude fiber and 11.5% protein and the leaves contained 17.1% crude fiber and 25.5% protein. Leaves must be preserved if forage is to be of high quality. Leaf shattering losses increase markedly when forage is handled at low moisture. When forage is handled at high moisture, field losses are often inversely related to storage losses since excessive moisture can cause serious problems during storage. Provided proper storage or curing facilities are available, total field and storage losses may be kept to a minimum with handling and storage at medium moisture. Shattering losses of alfalfa leaves due to raking are illustrated in Fig. 9.3. In this example, losses increase almost exponentially as the moisture content of the hay declines. The same type of curve would be representative of field losses incurred in other handling operations, such as stacking or baling at low moisture.

Windrower machines have eliminated most of the raking operations, so in drier climates they have reduced field losses. Since hay dries slower in the windrow than in the swath, windrowing prolongs field curing time; but the risk from weather damage and subsequent handling operations becomes greater. Under humid conditions, windrowing may have no advantage in reducing losses. Any vigorous handling process, such as raking or tedding (fluffing), can cause marked leaf loss if carried out at a moisture content much less than 50%.

Shrock and Fairbanks (1975) compared yield, moisture, and crude protein content of alfalfa hay packaged by a large roll baler, a compressed stack machine, and a conventional baler. The Kansas researchers concluded that when operating in dry, shatter-prone alfalfa hay, dry matter losses may be higher with either the stacker or the roll baler than with the conventional baler. When hay was packaged under desirable moisture conditions, no differences in dry matter yields occurred.

Crushing or crimping (conditioning) freshly cut forage breaks the stems and allows them to dry at a rate more nearly equal to the leaves. As a result, the forage dries at a faster rate (Fig. 9.4). The reduced exposure time to weather due to faster drying of the conditioned forage is important in reducing forage losses. Conditioned hay is often removed from the field and stored 1 day earlier than non-conditioned hay. However, if it rains, more moisture may be absorbed by conditioned hay than by non-conditioned hay. Conditioning is generally much more effective with legumes than grasses. Hundtoft (1965a) has compared dry matter losses among systems of handling forage (Table 9.1). These ranges are often much wider because of storage conditions. Dale et al. (1978) in Indiana developed a computer model entitled HAYLOSS which calculates both metabolic and mechanical losses on an hourly basis from the time of mowing through the baling process.

QUALITY CHANGES IN HAY DURING STORAGE

Dry hay may be stored in loose stacks, bales, small bunches left in the field, and other ways. Usually the climate and the storage package determine whether hay packages are left outside or are placed under some kind of shelter.

Nutrient Changes in Stored Hay

If hay is put into stacks or packaged at low moisture and protected from the weather, few nutrient losses occur during storage. However, due to oxidation, vitamin A loss is significant, and loss is greater on the outside of stacks or packages than toward the center. At high temperatures, vitamin a loss is greater than at low temperatures; thus, losses tend to be the greatest in warm weather. Carotene (precursor of vitamin A) losses of 50 to 75% after 1 year of storage are common regardless of the hay storage conditions. Under warm conditions with adequate oxygen, considerable loss of carotene may occur within 6 months. Absolute losses are greatest in hay with high initial content of carotene, such as in high quality alfalfa. Even though caro-

Table 9.1—Losses typical of forage handling systems.†

Moisture content when harvested	Field losses	Storage losses	Total
	%		
Direct cut silage (Above 70% moisture)	1-2	18-22	19-24
Wilted silage (65% moisture)	5-7	8-12	13-19
Haylage (sealed silo)	12-18	3-7	15-25
Heat-cured	12-18	3-4	15-22
Mow-finished (40% moisture)	12-18	4-6	16-24
Field-cured hay (20% moisture)	15-22	2-4	17-26

† From Hundtoft (1965a).

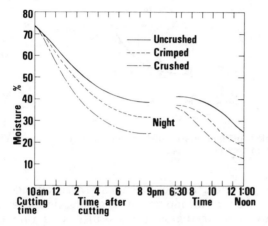

Fig. 9.4—General drying characteristics of alfalfa hay after conditioning. From Longhouse (1960).

tene oxidation takes place in dry stored hay, there is little change in energy, protein, or in any other nutrients. Hay stored inside is stable for years, although it may become brittle with age and be less palatable.

Large packages should be put up with no more than 20 to 25% moisture. Large packages refer to stacks made by a variety of machines and large round bales. Thirty percent moisture is generally too high for successful storage of large packages, except for grass hay in dry areas. Grass hay can be packaged at a slightly higher moisture than hay with a significant legume content. Storage losses in compressed alfalfa stacks in Ohio increased markedly when the moisture content exceeded 22.5% (Fig. 9.5). Hay with 25% moisture goes through a "sweat" which involves some heating as the excess moisture leaves the hay.

Large hay packages left outside are popular due to the reduction in labor requirements at harvest. In Indiana, Smith et al., (1974) suggested that if TDN loss due to weathering is no more than 10% based on the final weight of the package, a storage structure for large pack-

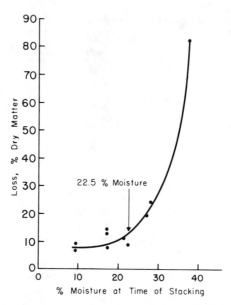

Fig. 9.5—Spoilage loss in compressed stacks made from alfalfa hay at different moisture contents. From drew et al. (1974).

ages cannot be justified. Large packages stored outside should not
touch each other, so that air movement is good. Air space between the
packages improves the drying-out process after each rain and reduces
the chance for mold damage and deterioration (Fig. 9.6). Weathering of
grass or alfalfa hay in the Indiana studies was generally confined to
the outer 5 to 10 cm of large bales and the top 10 to 15 cm of compressed stacks (Smith et al., 1974). The in vitro dry matter disappearance of weathered portions of large round bales of grass hay was 16
percentage points lower, and weathered mixed grass-alfalfa was 22
percentage points lower than the unweathered portions of the bales
(Lechtenberg et al., 1979).

The resistance of large packages to weather depends on how well
the packages are made. With compressed stacks, Indiana researchers
reported that an average of 68% (range 56 to 82%) of the forage remained unweathered after 1 year of outside storage and 59% (range 50
to 72%) remained unweathered after 2 years. With large round bales
73% (range 61 to 86%) was unweathered after 1 year and 70% (range
61 to 82%) remained unweathered after 2 years of outside storage. The
wide range in individual package values indicate that operator skill is
extremely important when making large hay packages (Parsons et al.,
1977). Losses tend to be higher with small round bales since the surface area is greater. Storage of large hay packages on a well-drained
site is important since a significant amount of weathering loss occurs
at the place of contact between the package and the soil. When large
round bales were stored outside from May until February on crushed
rock in Indiana, 85.5% of the original weight remained unweathered

Fig. 9.6—TOP: DO—Stack large packages so as to allow air movement around them so they can dry quickly after rain. Bottom: DON'T—Stack large packages tightly together outside or large losses from mold will occur where they touch. Photos by Lowell E. Moser.

compared to 76.9% for bales stored directly on the ground. Since 92% of the dry matter remained on bales stored inside, which were un-weathered, evidently 8% of the losses reflected in all the bales was dry weight loss during storage that was not related to weathering (Lechtenberg et al., 1979). Apparently, outside storage losses in large packages can be held to low levels in many temperate areas with moderate rainfall if a good package is made.

Feeding management of large hay packages is important. In Indiana, over 30% more alfalfa or grass hay was required for cattle that were fed large packages without feed racks compared to the amount required when a large package was fed on a concrete slab with a rack around it. Where an 8-day supply of forage was supplied to cattle, up to 50% more hay was required when fed without feed racks compared to feeding in racks (Smith et al., 1974).

In Alabama (Renoll et al., 1974) a square bale and a stack system were compared for johnsongrass [*Sorghum halepense* (L.) Pers.] hay handling. No differences in mineral content, crude protein, cell wall, or non-cell wall percentages were found between hay that had been stored in compressed stacks or bales. Digestibility for baled hay was higher than for stacked hay after 4 to 6 months of storage. Bales were covered with a tarpaulin and the stacks remained uncovered. Hay wastage at feeding was more of a problem with stacked than with baled hay. As the length of time to consume a stack increased and as the ground conditions became wetter, the hay losses from stacks increased markedly.

Heating of Wet Hay

Wet hay or dry silage (30 to 50% moisture) may heat and result in considerable nutrient loss. In normal situations, a rise in temperature to about 50 C should not be alarming since a temperature rise occurs in the "sweating" process. If the temperature reaches 60 C, the drop in feed value is of concern. The higher the temperature the higher the oxidative losses, and the most easily digested nutrients are oxidized first. Table 9.2 outlines the heating process in wet hay or dry silage. Heating is a greater problem in immature, high quality forages than it is in mature forages. Immature forage has more soluble nutrients that can support more respiration and bacterial action. If heated sufficiently, hay or silage darkens from a slight charring or carmelization of the sugars. Carmelization may actually make the hay a little more palatable, but the nutrient loss may be sizable. Overall digestibility, and especially protein digestibility, is lowered markedly by excessive heating due to the non-enzymatic browning reaction, also known as the Maillard reaction (Van Soest, 1965).

Table 9.2—Steps in heating of stored forage.

Temperature, C	Processes	Nutrient changes
Ambient–40	Normal "sweating" process, possible cell respiration, limited microbial action. Some fermentation can take place.	Excess moisture driven off. Very small respiration loss.
40–50	Normal "sweating", microbial action, plant processes stop at 45 °C. Some fermentation may take place.	Excess moisture continues to leave, very little loss.
50–60	Thermophylic microorganism activity. Non-enzymatic browning begins.	Lowered digestibility, lowered protein availability.
60–70	Thermophylic microorganism activity, increased oxidative reactions. Non-enzymatic browning continues.	Further lowering of digestibility and protein availability.
70–80	Biological activity ceases. Strictly chemical oxidative reactions. At 80 °C temperatures may rise very rapidly. Severe non-enzymatic browning, carmelization of sugars.	Very high losses in digestibility and protein availability.
80–280	Oxidative reactions occur rapidly due to high temperature.	Charring of forage. Large dry matter loss.
280–300	Oxidative reactions continue.	Possible ignition if ample oxygen is present.

Non-Enzymatic Browning

As early as 1909 Kellner reported that the digestibility of protein was 86.5% when grass hay was light brown in color, 75.1% when it was dark brown, and only 2.6% when hay was black. We refer to the reaction that causes lowered overall digestibility and lowered protein digestibility as non-enzymatic browning or the Maillard reaction. Heated forage is dark in color and the darkness is closely associated with the degree of heating and to the nutrient loss.

Heated forages show an actual increase in lignin and fiber. The increase in acid detergent fiber (ADF) (cellulose and lignin) can be accounted for largely by the formation of artifact lignin in the nonenzymatic browning process. A N analysis of the ADF fraction will detect the amount of bound N that would be unavailable to the animal. If ADF N is subtracted from total N and multiplied by 6.25, the percentage of available crude protein can be calculated. The non-enzymatic browning reaction involves condensation of carbohydrates with free amino groups of proteins or amino acids, which forms dark-colored, indigestible polymers that contain N. Sugar residues appear to condense with amino groups at a 1:1 ratio. Carbohydrates, lipids, or proteins may also condense with lignin or with each other. Both highly digestible carbohydrates and protein are lost to the animal. This reaction becomes especially significant above 55 C. Formation of artifact lignin is highest at high temperatures, with high pH, and in plants with high sugar and protein content. Water is required in the reaction and non-enzymatic browning occurs at a maximum in forage at 30% moisture (wet hay) (Van Soest, 1965). Heated forages will be of lower digestibility, and extra protein may have to be supplemented to meet livestock needs.

Effect of Mold on Hay Value

If oxygen is present, and temperatures do not become excessively high, molding may occur and markedly reduce feed value of hay. Saprophytic fungi are present on all forage and if sufficient moisture and oxygen are present during storage they will grow and develop. Such organisms will use the readily available carbohydrates and thus lower feeding value. Protein in the forage is degraded and ammoniacal N is released. In addition, moldy hay has been associated with lowered livestock intake and lowered animal performance. Cattle are fairly tolerant of moldy hay provided it is not their total diet. Some molds produce mycotoxins, which are toxic to livestock. The mycotoxin problem does not occur often, but moldy hay should be fed with caution.

Hay Preservatives

In Humid climates field curing hay dry enough to store often is difficult. For years, common salt (NaCl) has been thought to act as a preservative for wet hay. Salt may function as a control of the microbial action and as a result may prevent hay from heating excessively. In many experiments salt has delayed rather than prevented heating of undercured hay. However, in some instances it has reduced mold. If salt is to be effective, 1 to 2% must be applied. Feeding problems have been experienced with this high rate of salt. Applying high amounts of sodium on soils through livestock wastes may cause soil dispersion.

The successful use of organic acids as preservatives on wet corn (Zea mays L.) prompted investigations on their use for preventing nutrient loss in heating and molding of undercured hay. Propionic acid has strong fungicidal and fungistatic properties. In Indiana, hay baled at 32% moisture and treated with 1% propionic acid did not mold or heat, and dry matter loss was reduced from 15 to 7.6% (Knapp et al., 1976). Digestibility decreased in both the hay treated with 1% propionic acid and the untreated hay. The treated hay was higher in digestibility than the untreated hay. Retaining higher digestibility appeared to be due to a higher carbohydrate level since the propionic acid did not increase cell wall digestibility.

Anhydrous ammonia, due to its fungicidal properties, also acts as a hay preservative. Experiments in Indiana have shown that anhydrous ammonia applied as a gas at 1% of the hay weight after baling hay at 32% moisture prevented heating, molding, and quality deterioration (Knapp et al., 1975). Dry matter and sucrose losses were reduced during storage. Crude protein, ammonium-N content, and cell wall digestibility significantly increased in treated compared to untreated hay. Concentration of other carbohydrates and dry matter digestibility were lower in the treated forage than in the original material, but were higher than in the untreated hay. Anhydrous ammonia not only acts as a preservative but may increase digestibility and add nonprotein N to the hay. Practical ways of application to large volumes of hay still present a problem.

PROCESSING FORAGES BEFORE FEEDING

Changing the physical form of hay can change the value of a forage for livestock production. Hay may be ground to increase livestock intake of low quality forage, to reduce livestock rejection of the more unpalatable portions of the plant, or to make possible the blending of a ration. Grinding of hay may reduce its digestibility per se because the rate of passage through the rumen increases. The digestibility of fiber decreases more than the soluble fraction. Finely ground hay moves through the rumen faster than more coarsely ground hay. When forage passes rapidly, it is exposed to the rumen microflora for a shorter period of time. Animal performance is often better with low quality forages when they are ground compared to when they are fed unprocessed because the rumen is not filled with a low digestible forage that is moving through slowly. The difference is not nearly as great with high quality forage. A rapid rate of passage improves forage intake, and the animal has access to more nutrients. Forage must be finely ground to 1 cm or shorter in length in order to speed up the rate of passage through the rumen. Forage may be chopped or coarsely ground to 1.0 to 5.0 cm in order to incorporate it into a ration, facilitate handling, or to reduce loss due to animal refusal. If ground too finely, forage becomes dusty and unpalatable. Adding 1% tallow, water at grinding, or molasses to the ground ration can overcome much of the dustiness (Hale and Theurer, 1972).

Pelleting and Wafering

Pellets, wafers, and cubes differ in their physical form (Fig. 9.7). Pellets are the most dense and wafers are the least dense. Cubes are the most dense form of forage package where grinding is not done prior to packaging. After fine grinding, pelleting may further improve consumption of low quality forage and make finely ground forage more palatable. It may also make ground forage easier to handle or transport. Increased consumption of high quality forage may result from grinding or pelleting, but not always. Generally, pelleting the forage will increase consumption, increase livestock production, and reduce

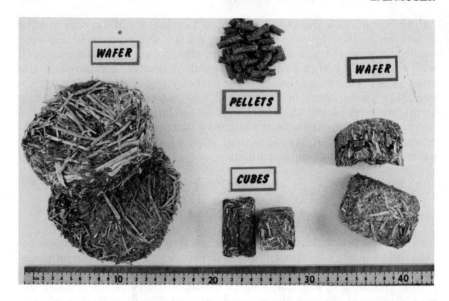

Fig. 9.7—Forage pellets, wafers, and cubes differ in their physical form. A centimeter ruler is at the bottom of the picture. Photograph from the collection of Kenneth Von Bargen; photo by Bruce Sandhorst.

the feed requirement per unit of livestock production. Since grinding and pelleting increases volatile fatty acids, especially propionic acid, in the rumen, feeding pelleted forages ad libitum to dairy animals may increase milk production but often decreases milk fat markedly.

Wafers are made from chopped rather than ground hay. Since wafering does not change the physical form as drastically as grinding and pelleting, the feeding value of wafered hay differs little from hay in other long forms.

Cubing compresses hay into small cubes and is a modification of wafering. Cubing has been limited generally to pure alfalfa grown under irrigation in dry climates where windrow drying down to 10% moisture content is possible without excessive field losses. Water is sprayed on the windrow to raise the moisture back to about 14% so that a stable cube can be formed. Leaf losses are quite low with successful cubing. Cubes can also be made by stationary units. Cubes are fed without further processing except that they often are crumbled when fed to young animals. Cattle fed cubes usually consume more forage than those fed long hay, but not as much as they would consume from pelleted forage. Cubes are used in large-scale dairy operations in climates suitable for cubing. Cubes are fairly dense, can be easily mechanized, and are easier to ship than hay in long form. However, the cubing process requires more energy than packaging hay in other long forms.

Fig. 9.8—Rapid artificial drying of alfalfa results in a high quality dehydrated forage (Dehy) that may be used as a protein, vitamin, and general supplement in livestock rations. Photo courtesy of American Dehydrators Association.

Dehydrated Forages

Dehydrated forages, primarily alfalfa, have been used since the 1930's to supply protein for livestock along with vitamins A, E, and K, xanthophylls, and "unidentified growth factors" for monogastric animals, particularly poultry (Fig. 9.8). The basic objective in dehydration is to remove water as rapidly as possible from forage thus keeping losses of nutrients low. Dehydration allows fast drying leaves to move through the drier quicker than slow drying stems. The evaporating moisture keeps the forage cool, and as long as the forage is not heated excessively after it has been dried, nutrients are not tied up. In the past in the USA, most of the alfalfa used for dehydration has been cut while standing and taken directly to the dehydrating drum. Such direct-cut material contains 70 to 80% moisture. Recently, field wilting to around 60% moisture has reduced the energy required for drying 10 to 30%; however, the energy for the harvesting operation is increased. The only loss incurred in wilting to 60% moisture is a small loss of carotene and xanthophyll, but the protein remains about the same. Possibly in the future mechanical removal of water from alfalfa will be used before dehydration (Smith, 1977). In dehydrating and subsequent processing drying, grinding, and storage losses can all occur.

Fig. 9.9—Storage of dehydrated alfalfa pellets under carbon dioxide gas in tanks such as these reduces storage losses to near zero. Photo courtesy of American Dehydrators Association.

In a well managed dehydrating plant, drying and grinding may cause a 10% loss of carotene; generally less than 10% of carotene is lost in the field before drying. A 20 to 70% loss of xanthophyll may occur during dehydration. Xanthophylls are desirable since they increase the pigmentation of poultry and egg yolks. Lysine losses range from 15 to 50%. Protein content and digestibility are maintained very well with rapid dehydration compared to other ways of processing forages. The moisture level in the forage at the discharge end of a drier should not be lower than 8 to 10% because lower moisture will cause high overall losses (Kohler et al., 1972). Vitamin stability, during storage, and dustiness have been largely controlled in the dehydration industry.

After dehydrated alfalfa is ground, it is pelleted to overcome the dustiness and to facilitate handling. Oxygen-free storage of the alfalfa pellets in a CO_2 atmosphere has essentially eliminated oxidative losses of vitamins and other compounds. Of course, insects, rodents, or heating losses do not occur in a CO_2 atmosphere (Fig. 9.9). Since oxidative losses occur after alfalfa pellets are removed from CO_2 storage, an antioxidant, usually a compound called ethoxyquin, is applied at about 0.015% during processing to reduce losses of carotene (vitamin A), xanthophyll, and tocopherol (vitamin E). Vitamin losses with antioxidants are only 35 to 50% as great as they are with untreated alfalfa pellets.

SILAGE

Crude methods of ensiling (anaerobic fermentation of forages) have been attempted since ancient times and by the mid-19th century ensiling forage crops was common in most of Europe. By late in the 19th century ensiling had spread to other temperate forage-producing regions of the world.

Ensiling forages offers a way to overcome problems in drying hay in humid forage-producing areas. Mechanization of silage harvest and feeding has been easier to accomplish than mechanization of hay harvest. Many crops ensile easily and field losses are often lower for silage than for hay. For these and other reasons, ensiling is popular in most forage-producing regions where the storage of forage is necessary.

Chemistry and Microbiology of the Ensiling Process

The chemistry and microbial changes during ensiling are complex and varied due to the many types of forages that may be ensiled and the great differences in the composition of forage feeds. More problems occur in ensiling grass and/or legume silage than in ensiling grain crops since moisture and protein content are often high and carbohydrates are low. Silages containing high amounts of fermentable carbohydrates, such as corn, sorghum [*Sorghum bicolor* (L.) Moench.] and other high energy silages, have simpler chemistry and microbiology. Although some losses during ensiling are unavoidable good management minimizes overall losses.

Carbohydrates, Proteins, and Vitamins

The first stage in ensiling is aerobic (Table 9.3). Some plant cells will be alive and plant enzymes still active. In addition, a wide assortment of microflora exists on the forage. Aerobic coliform and pigmented bacteria are present in relatively large numbers, but die quickly under the anaerobic conditions. Aerobic respiration may result in the loss of some sugars, but since the aerobic stage should last for only a few hours, these losses are not great. The second stage involves anaerobic bacterial action and the fermentation reactions of the ensiling process.

Very few obligate anaerobes (organisms that function only under anaerobic conditions), which produce the majority of the organic acids in silage, are present on freshly chopped material. They do not increase rapidly in numbers until several days after ensiling. Few of the organ-

Table 9.3—The ensiling process.

Time	Environment	Bacteria involved	Chemical changes
1 day	Stage I **Aerobic Stage,** increase in temperature, depletion of O_2.	Aerobic coliform and pigmented bacteria.	Aerobic plant cell respiration. CO_2 production, carbohydrate utilization, some protein hydrolyzed.
1 day to 3 weeks	Stage II **Anaerobic Stage,** fermentation, pH continuously decreasing.	Lactic acid producing bacteria. Depending on crop, many strains of bacteria can be isolated (e.g. *Lactobacillus*)	Carbohydrates converted to lactic, acetic, succinic acid, etc. Protein hydrolysis to amino acids. Some hemicellulose breakdown. pH drops significantly.
After 3 weeks	Stage III **Post-fermentation Storage Stage,** should be relatively stable, have low pH and a pickled smell. May involve pH rise and spoilage if *Clostridium* or other sporeformers are active.	Bacterial action stopped unless pH is not low enough, then *Clostridium* and other spore-formers are active.	If bacterial action stopped, very little change-stable storage. If *Clostridium* are active, lactic acid converted to butyric and propionic acids. Amino acids deaminated and decarboxylated forming ammonia and foul smelling amines.

isms on green plants are similar to bacteria found in silage and fewer still are typical of *Lactobacillus,* the lactic acid forming bacteria. Obligate anaerobes produce primarily lactic acid which lowers the pH of the silage to a point where further bacterial action is checked. In silage, bacteria that form less acid are gradually replaced by those that form more acid until all bacterial action ceases. In grass silage the coliform group is replaced by the spherical lactic-acid producers which, in turn, are replaced by strongly acidifying, rod-shaped, lactic acid bacteria (Langston et al., 1958). The buffering capacity of forage resists the lowering of pH, which stops bacterial action. Legumes are more highly buffered than grasses and, consequently, with direct-cut legumes, there are often more ensiling problems. The increased buffering activity is associated with the high protein content which is partially hydrolyzed to amino acids, as well as the cations in legumes, although there are many other types of buffers in plants. A third stage in the ensiling process may result when undesirable spore-forming anaerobes such as *Clostridium* form butyric acid and degrade protein if excessive moisture is present.

    Carbohydrate fermentation may occur by two pathways (Whittenbury et al., 1967) (Table 9.4). Fructosan and sucrose are quickly broken down to fructose and glucose, and the two hexoses may be metabolized differently depending on the organisms involved. Homofermentative lactic acid bacteria ferment glucose or fructose to two molecules of lactic acid each. With glucose, heterofermentative lactic acid bacteria produce one molecule of lactic acid, one of ethanol, and one of CO_2. With fructose, heterofermentative bacteria produce mannitol and acetic acid in place of the ethanol. Less lactic acid is produced by heterofermentative organisms acting on fructose than on glucose, so the fructose fermentation process is less efficient if a significant amount of heterofermentative organisms are present. Pentoses are fer-

Table 9.4—Products of carbohydrate fermentation in silage.†

Homofermentative reactions

a) 1 glucose ⟶ 2 lactic acid
b) 1 fructose ⟶ 2 lactic acid
c) 1 pentose ⟶ 1 lactic acid + 1 acetic acid

Heterofermentative organisms

a) 1 glucose ⟶ 1 lactic + 1 ethanol + 1 CO_2
b) 3 fructose ⟶ 1 lactic acid + 2 mannitol + 1 acetic acid + 1 CO_2
c) 1 pentose ⟶ 1 lactic acid + 1 acetic acid

† Whittenbury et al. (1967).

mented by both homo and heterofermentative organisms to lactic and acetic acids. If the heterotypes predominate, a lower productivity of desirable acid will result. The homotypes predominate in good quality silages according to Langston et al., 1958. Organic acids in plants can also be fermented to yield simpler acids and CO_2. Cellulose is relatively unaffected by the microbial activity of the ensiling process. The hemicelluloses to a certain extent may be hydrolyzed, which produces pentoses for fermentation. In unwilted forage 10 to 55% of the hemicellulose is apparently hydrolyzed during ensiling.

Protein hydrolysis also occurs in the ensiling process. About 80 to 90% of the N is in the protein form before ensiling and about half of the protein is hydrolyzed to amino acids during the ensiling process (Whittenbury et al., 1967). Protein hydrolysis does not necessarily cause a N loss, although amino acids may be deaminated or decarboxylated by *Clostridium* bacteria if the pH is not low enough. The true protein content (amino acids still connected in protein chains) may be very low in a silage sample but the crude protein (N × 6.25) may be as high as it was in the original forage. The N bases also cause silage to have a high buffering capacity.

Vitamin content is maintained fairly well under anaerobic silage conditions. Carotene (vitamin A) preservation in silage is better than it is in hay and the same is true for most other vitamins. Considerable vitamin C (ascorbic acid) in silage is unavoidably lost. The B vitamins and vitamin D change little.

Toxic Compounds

Toxic compounds in forages are often detoxified during the ensiling process. Since protein is denatured and some is hydrolyzed, bloat is usually not a feeding problem. Ensiling is recognized as a way to reduce high nitrate in forages. Most of the time, after 90 days of storage, silages contain little nitrate unless high levels were originally present. When high nitrate materials are ensiled, however, toxic levels of nitrate may remain at feeding time. Nitrate disappearance is reduced by a low pH. With the high pH fermentation that occurred with high moisture silages, nearly all of the nitrate was removed. With wilted silage, less nitrate (53 to 55%) was removed because microbial activity was much less. With low moisture silage, only 19 to 21% of the nitrate disappeared (Jacobson and Wiseman, 1963). Nitrate removal appears to be greatest with maximum microbial activity. If microorganism activity is checked by low pH or lack of moisture, nitrate removal will be reduced. Any additive that hastens the drop in pH will also reduce nitrate removal. Possibly nitrate removal takes place in two stages during the fermentation process. Some disappearance occurs shortly after the silage is placed into the silo and before the pH is lowered. Laceration, which occurs when chopping silage, hastens nitrate disappearance before the pH is lowered and complete nitrate removal may be possible. More nitrate can disappear later if for some reason the pH becomes elevated.

Cyanogenic glucosides (prussic acid) are detoxified during ensiling in a fashion similar to that described for drying. Therefore, poisoning from cyanogenic glucoside-containing plants, such as sudangrass, generally does not occur with silage.

pH-Acid Content

Few mineral nutrients are lost during normal ensiling, although they may change form. For example, even excellent silage loses its green color in the silo. This change is due to the magnesium atom being removed from the chlorophyll molecule by organic acids. The magnesium-free chlorophyll derivative is brown.

If forage is excessively wet upon ensiling, poor fermentation results and the action of *Clostridium* bacteria produces butyric acid from the lactic acid, deaminates amino acids to form ammonia, and decarboxylates amino acids, forming amines. Some of the amines, such as putrescine and cadaverine, are foul smelling and give wet grass silage an obnoxious odor. Low quality silage is associated with, but not necessarily due to, high levels of butyric acid. Action of *Clostridium* can be inhibited by allowing forage to wilt to a moisture content below 70% or to increase the acidity of the silage. *Clostridium* is especially sensitive to the high osmotic pressure created by wilting forages; therefore bacterial activity in wilted silages is much less than that in high moisture silage. *Clostridium* tolerates high concentrations of organic acids if excess water is present. The wetter the material, the lower the pH must be for preservation. With wet material, natural fermentation is not enough to lower the pH sufficiently. Wilting allows the plant sap to concentrate, resulting in faster fermentation and lowering of the pH below a satisfactory level where *Clostridium* functions. *Lactobacillus* works well at a pH where the activity of *Clostridium* and other sporeformers is stopped. Temperature of the silage has a great effect on the undesirable spore-forming bacteria. Ten times as many spore-forming bacteria were found when the temperature was 28 C compared to 16 C (Langston et al., 1958).

Good quality grass and legume silage has a pH of 3.9 to 4.8, lactic acid on a dry weight basis from 3 to 13%, ammonium N on a dry weight basis of 1 to 3%, none or only traces of butyric acid, and low

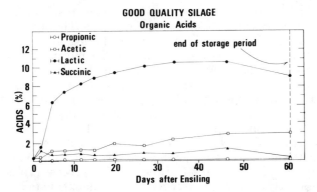

Fig. 9.10—Organic acid level in typical good quality grass silage. From Langston et al. (1958).

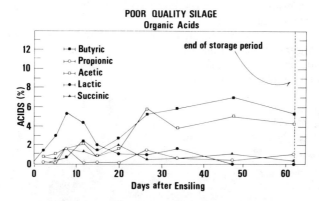

Fig. 9.11—Organic acid level in typical poor quality grass silage. From Langston et al. (1958).

spore counts (Fig. 9.10) (Langston et al., 1958). Poor quality silage has a pH of 5.2 to 5.7, low lactic acid, ammonium N greater than 3.2%, up to 7.5% butyric acid, and high spore counts. Lactic acid content may be high in low quality silages early in the ensiling process, but after action by anaerobic spore-forming bacteria, lactic acid may be low (Fig. 9.11).

Unwilted silage is low in carbohydrates, high in volatile and non-volatile organic acids, low in true protein, and high in non-protein N compared with the original ensiled material. Fiber and mineral content are about the same.

Low Moisture Silages

As discussed previously, moisture content may change the direction of the ensiling process by not allowing pH to drop below the level which *Clostridium* activity will be stopped. Wilting to 60 to 70% moisture overcomes this problem. In the late 1950's and early 1960's "haylage" was introduced. Haylage, or low moisture silage, results when grasses and/or legumes are wilted to 40 to 60% moisture. The chemistry of low moisture silage is much different, and excluding air becomes much more of a problem, than with wilted (60 to 70% moisture) or direct cut (>70% moisture) silages. With low moisture, less bacterial fermentation occurs and thus the pH often remains greater than 5.0. Special care, such as fine chopping, packing, and sealing, must be taken at storage time to exclude air. Although fermentative action is low, silage around 50% moisture is particularly prone to heating and molding if oxygen is present. The Maillard reaction, which has been discussed in a previous section, may cause haylage to be of low quality. Molds and yeasts will grow during Stage I of the ensiling process (Table 9.3) but should die quickly when the oxygen supply is exhausted. If aerobic conditions are extended, the silage becomes moldy and has very little feed value. Silo fires have occurred where the moisture content was 25 to 40%. Ammonia can often be detected in moldy silage, and then loss of protein is especially high.

The type of silo structure greatly affects the losses of low moisture silages. In general, low moisture silage should not be put in bunker or trench silos or stacked on the ground unless they are carefully covered with higher moisture material (Fig. 9.12 and 9.13).

Fig. 9.12—Upright silos vary from concrete to glass lined. Storage losses are generally lowest in this type of silo compared to bunker or trench silos. Photo by Lowell E. Moser.

Losses in Silage

Silage losses can be classified as avoidable and unavoidable. Unavoidable losses include plant respiration and normal fermentation losses. Avoidable losses may include those from harvesting, mold, putrification, heating, and seepage. Losses depend upon the moisture content at ensiling and the type of storage structure used. Realistic estimates of losses of silage dry matter in regard to moisture content at ensiling are given by Shepherd et al. (1953) in Table 9.5.

Field losses increase as the material is field dried (Fig. 9.14). Losses are low in dry (40 to 60% moisture) silages in sealed silos where little fermentation occurs. Losses may be high with dry material in structures where air cannot be excluded. Surface losses are generally less with upright silos since the exposed surface is relatively small. However, seepage losses are greatest in upright type silos since the pressure per unit area is greater than with a low silage stack. Seepage is important since soluble carbohydrates, soluble proteins, and organic acids are all nearly 100% digestible and should be retained in the silage. Seepage delays the pH drop, which is necessary to check harmful microbial activity. If rain enters the silage, leaching can further contribute to seepage losses. Nonpermeable covers like polyethylene

Fig. 9.13—Losses are greatest in stacked silage, especially if low moisture silage is stored in small piles as shown here. Photo by Lowell E. Moser.

Table 9.5—Estimate of minimum dry matter losses in forage stored as silage at different moisture levels based on 6 months of storage.†

Kind of silo, and moisture content of forage as stored	Surface spoilage	Fermen- tation	Seepage	Total silo losses	Field losses	From cutting of crop to feeding
			%			
Conventional tower silos:						
85%	3	10	10	23	2	25
75%	3	8	3	14	2	16
65%	4	8	0	12	4	16
Gas-tight tower silos:						
85%	0	10	10	20	2	22
75%	0	8	3	11	2	13
65%	0	6	0	6	4	10
50%	0	4	0	4	10	14
Trench silos:						
85%	6	11	10	27	2	29
75%	8	9	3	18	2	20
70%	10	10	1	21	2	23
Stack silos:						
85%	12	12	10	34	2	36
75%	16	11	3	30	2	32
70%	20	12	1	33	2	35

† From Shepherd et al. (1953).

reduce silo losses greatly. In New Zealand, total dry matter loss in small bunker silos was only 12% when weighted polyethylene was used as a cover compared to 34% loss with no covers (Minson and Lancaster, 1965) (Fig. 9.15).

Various types of forage harvesting systems cause differences in overall losses (Moore, 1951) (Fig. 9.16 and 9.17). Forages handled in dry form have more field loss. Wilted silage has the advantage of reducing exposure time and keeping field losses low. However, fermentation and other ensiling losses occur with increasing moisture. An overall view of field and storage losses in relation to moisture content is shown in Fig. 9.14.

Preservatives and Additives

Since ideal storage conditions and harvesting at ideal moisture levels are not always possible under farm conditions, the use of preservatives or additives may be a useful tool. In situations where the moisture is below 70% and silages are high in fermentable carbohydrates, preservatives are generally not needed to insure a desirable ensiling process. Various types of additives may be used to enrich the silage in a particular nutrient, and such additives may affect the

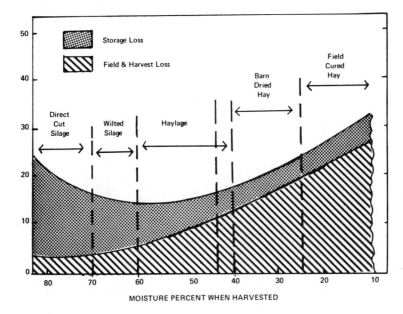

Fig. 9.14—Estimated total harvest loss and storage loss when legume-grass forages are harvested at varying moisture levels and by alternative harvesting methods. From Hoglund (1964).

Fig. 9.15—Weighted polyethylene makes an effective cover and reduces losses in trench and bunker silos or in silage stacks. Photo by Lowell E. Moser.

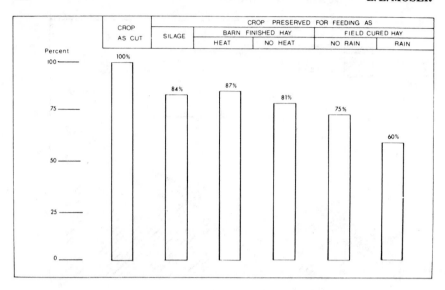

Fig. 9.16—Dry matter preservation under various forage handling systems. From Moore (1951).

fermentation process too. Preservatives may act in a number of ways to alter the ensiling process. They may add fermentable carbohydrates, restrict bacterial action, lower pH of the silage, or stimulate lactic acid formation.

Direct acidification of high moisture silage to lower the pH below the level which the undesirable *Clostridium* bacteria can act on the silage has been successful. A. I. Virtanen in Finland first introduced the method of adding mineral acids such as dilute H_2SO_4 and HCl to wet silages to obtain a pH of less than 4.0. Numerous studies have shown mineral acids to be an effective way to preserve silages resulting in less protein breakdown, and lowered respiration and fermentation losses. The AIV method, as it is called, has not been used widely throughout the world because of the difficulty in handling strong corrosive mineral acids. Problems have also occurred with livestock intake of AIV silages. Organic acids such as lactic, acetic, and more recently, formic acid, have been added to silages. They act as direct acidifiers, stopping undesirable microbial action. Results with formic acid have shown it to be an effective preservative and it has improved feeding value of direct-cut silage. Acid salts such as calcium formate and sodium nitrite have been mixed and sold under commercial names. Results with these salts are inconsistent.

Sterilants that stop nearly all bacterial action, including lactic acid fermentation, have been used. Sodium metabisulfite is an example of a sterilant, and at a rate of 4 kg per metric ton (8 lb/ton) has often produced silage with better color and aroma than that of untreated silages. Protein degradation has been reduced. Adding sodium metabisulfite at 0.4% has caused no major problems, but at higher amounts silage intake has been lowered.

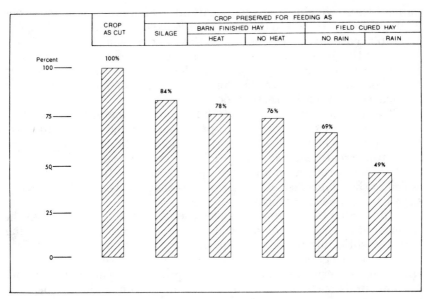

Fig. 9.17—Protein preservation under various forage handling systems. From Moore (1951).

Antibiotics to stop bacterial action have shown variable results as well. Bacitracin, the most common antibiotic used, has reduced protein loss, increased digestibility, and improved livestock performance compared to untreated silages. However, results cannot always be repeated.

Lactic acid stimulants include inoculation with lactic acid bacteria cultures and other organisms. Tremendous variation in results is evident from study to study, depending on the interaction of the organisms with the environment. In some cases, nutrient preservation has been better in silages inoculated with lactic bacterial cultures. In Nebraska, *Aspergillus* and lactic bacteria improved protein preservation, dry matter preservation, and feed efficiency when added to direct-cut alfalfa, but the mold-bacteria combination has little effect on corn silage (Owen, 1971). The enzyme cellulase has been used to break down cellulose to obtain more carbohydrates for lactic acid fermentation. Limited success has been achieved. Sugar and molasses may be added to low carbohydrate silages to provide additional substrate for bacterial fermentation. Addition of 1 to 2% molasses or 1% sucrose resulted in faster fermentation. Dried whey, a source of lactose, acts in much the same way when applied in the range of 1 to 3%. Liquid whey is rather dilute for use as a silage preservative.

Other fermentable feedstuffs have also been used successfully to increase lactic acid production in silages low in carbohydrates. Ground corn, wheat (*Triticum aestivum* L.), potatoes (*Solanum tuberosum* L.), sugarbeet (*Beta vulgaris* L.) pulp, and citrus pulp are examples. The latter two are also effective in absorbing water from high moisture forages, thus reducing seepage. About 50 to 80 kg of ground grain per metric ton (100 to 160 lb/ton) of silage is used. Of course, adding fermentable carbohydrates increases the energy value of the silage. About 20 to 25% of the feed value of an added ground grain is lost in the silo due to various reactions. This loss must be considered as the "cost" of the preservative. The loss figure could be higher with sugar or molasses if excessive seepage occurs.

The addition of 0.5 to 1.0% limestone to material at ensiling neutralizes some of the lactic acid produced and allows more to form. If limestone is kept below 1%, pH can be maintained at 4.5 or less. Limestone treatment has been used to increase feed efficiency in regard to beef production with high energy silage. In addition, Ca is added to the silage. As with many of the other additives, consistent results have not been obtained.

Adding 0.5 to 1% urea increases the nonprotein N fraction in the silage, and neutralizes some of the lactic acid. Urea increases ensiling losses slightly due to increased fermentation and gaseous loss. Urea and limestone added in combination act similarly to each added alone. Nitrogen recovery from added urea is approximately 80 to 90%. Other N compounds such as biuret, ammonia, and diammonium phosphate have been used to enhance N content of high energy silages. An ammonia-mineral-molasses mixture, developed in Michigan, increases lactic acid and adds N. The ammonia is retained as salts of the organic acids.

Hay crop silages wilted below 70% moisture generally will not benefit from preservatives, especially if they are high in carbohydrates. Research results have shown that high protein silages with moisture contents greater than 70% will benefit if the preservative will either directly or indirectly cause the pH to be lowered below 4.5 more quickly. A fast pH drop will limit putrifying organisms such as *Clostridium*.

The whole area of silage preservatives for the producer has been clouded by inconsistent research results and by people selling "magic potions" to greatly improve silage quality. Many good silage preservatives and additives are on the market, but the producer should be sure he really needs one, and if so, it should be selected on sound chemical and biological principles. There are no miracle-producing additives.

Livestock Utilization of Silage

Use of grass and legume silage by livestock, compared to the same crop fed them as hay, depends largely on the moisture content of the silages and the ensiling conditions. Livestock performance on direct-cut high moisture silage often is not as good as with hay or drier silage. The greatest differences occur where the silage makes up all or nearly all of the ration. Reduced performance with wet silage is associated with lowered feed intake rather than with lowered digestibility. Although reasons for lower intake of high moisture silages are unclear, they are probably not the amount of water per se, but may be due to compounds produced by fermentation under wet conditions. Excessively high levels and acids and low pH possibly reduce intake. Low moisture silages have proved advantageous with high producing animals where a high forage intake is important.

CHEMICAL TREATMENT OF LOW QUALITY ROUGHAGES

Vast amounts of grain crop residues and other low quality roughages exist in the world. Treatment to delignify residues and make them into a much higher quality feed for ruminant animals has been investigated. Crop residues in Nebraska, treated with 4% NaOH with subsequent "ensiling" after bringing the moisture content to about 60%, had the energy feeding value of about 90% that of corn silage (Klopfenstein, 1973; and Rounds et al., 1975). Grain sorghum and corn residues collected behind the combine appear to have the most potential for chemical treatment. Since Na causes a soil dispersion problem when disposing of manure and the chemical cost must be kept at a minimum, the use of at least some inexpensive $Ca(OH)_2$ might be considered. In Nebraska, substituting up to one-half $Ca(OH)_2$ for NaOH tended to increase or at least did not decrease rate and efficiency of gain. Calcium hydroxide would also add some Ca to the ration. Crop residues treated with 1% $Ca(OH)_2$ and 3% NaOH have produced the best gain and feed efficiency (Rounds et al., 1975). Ammonium hydroxide has been less successful but would add non-protein N to the feed as well. Forage treated with NH_4OH may have to be blended with normal silage to be acceptable. High pressure steam treatment (17.6 kg/cm^2 (250 lb/in^2) for 50 sec) of corn cobs was an effective way to increase feeding value of low quality roughages (Klopfenstein, 1975).

In vitro digestibility of rice straw was improved with a steam treatment at 160 C. Digestibility was increased more with high moisture straw than with low-moisture straw. With 2.6 or 5.2% of ammonia added before the steam treatment the rice straw digestibility increased markedly over the straw that was steam-treated only. An increase in digestibility was also obtained at room temperature and pressure if the ammonia was allowed to act on the straw for about 20 days (Waiss et al., 1972). Chemical treatment and subsequent "ensiling" of low quality forages may offer a way to expand forage feed supply in the future, depending upon grain and forage prices.

SUMMARY

Maintaining forage quality during harvesting and storage requires important management decisions. When dealing with grain crops, field or storage losses are of great concern. Equal concern should be expressed when leaf loss, moldy hay, silo seepage, or any other type of loss is observed in forages.

Since leaves contain most of the nutrients they must be preserved during harvest. Handling losses in dry forages are difficult to avoid. After dry hay is stored losses may be low if it is protected from the weather, but if stored outside the losses depend on climatic factors and ability of the package to shed water.

Silage losses are low at harvest but fermentation loss is unavoidable. Losses can be kept at a minimum if water content is not excessively high or low, and if oxygen is kept out of the silage mass. Preservatives might be needed if the moisture content is greater than 70% but the choice of a preservative should be based on sound reasons or biological principles, not a series of unfounded promises.

Forage is not usually improved during storage. However, some processing methods improve the feeding value of low quality forage. Grinding low quality forage can increase its feeding value. Chemically treating low quality forages like crop residues can potentially improve their feeding value.

Reducing harvesting and storage losses are part of the good management that is needed to insure adequate quantity and quality of forage at the time of feeding.

LITERATURE CITED

Burzlaff, D. F., and D. C. Clanton. 1971. The production of upland hay in the Sandhills of Nebraska. Nebraska Agric. Exp. Stn. Bull. 517.

Dale, J. G., D. A. Holt, R. M. Peart. 1978. A model of alfalfa harvest and loss. Paper No. 78-5030. Summer meeting, Am. Soc. Agric. Engineers, 27–30 June 1978, Logan, UT.

Drew, L. O., H. M. Keener, R. W. Van Keuren, H. R. Conrad, and W. E. Gill. 1974. Preservatives for hay? Ohio Rep. Res. Develop. 59(2):38–39.

Greenhill, W. L. 1959. The respiration drift of harvested pasture plants during drying. J. Sci. Food Agric. 10:495–501.

Hale, W. H., and C. Brent Theurer. 1972. Feed preparation and processing. *In* D. C. Church (ed.) Digestive physiology and nutrition of ruminants. Vol. 3—Practical nutrition. D. C. Church, Dep. of Animal Science, Oregon State Univ., Corvallis. Distributed by O and B Books, Corvallis, OR 97330.

Hesse, W. H., and W. Kennedy. 1956. Factors causing errors in the determination of dry matter and nitrogen in forage crops. Agron. J. 48:204–207.

Hoglund, C. R. 1964. Comparative storage losses and feeding values of alfalfa and corn silage crops when harvested at different moisture levels and stored in gas-tight and conventional tower silos: an appraisal of research results. Michigan State Univ., Dep. of Agric. Econ. Mimeo 946.

Hundtoft, E. B. 1965a. Handling hay crops—capacity, quality, losses, power, cost. Cornell Univ. Agric. Engineering Ext. Bull. 363. Ithaca, NY.

————. 1965b. Handling hay crops—from standing crop to windrow. Cornell Univ. Agric. Engineering Ext. Bull. 364. Ithaca, NY.

————. 1965c. The self-propelled windrower. Cornell Univ. Misc. Bull. 67. Ithaca, NY.

Jacobson, W. C., and H. G. Wiseman. 1963. Nitrate disappearance in silage. J. Dairy Sci. 46:617–618.

Kellner, O. 1909. The scientific feeding of animals. Duckworth and Co., London.

Klopfenstein, T. J. 1973. Processing and treatment of crop residues to increase utilization. Nebraska Crop Residue Symposium, 10–11 Sept. 1973.

————. 1975. Pressure treatment of corn cobs. p. 12–13. *In* 1975 Nebraska Agric. Exp. Stn. Beef Cattle Report. EC 75-218.

Knapp, W. R., D. A. Holt, and V. L. Lechtenberg. 1975. Hay preservation and quality improvement by anhydrous ammonia treatment. Agron. J. 67: 766–769.

————, ————, ————. 1976. Propionic acid as a hay preservative. Agron. J. 68:120–123.

————, ————, ————, and L. R. Vough. 1973. Diurnal variation in alfalfa (*Medicago sativa* L.) dry matter yield and overnight losses in harvested alfalfa forage. Agron. J. 65:413–417.

Kohler, G. O., E. M. Bickoff, and W. M. Beeson. 1972. Processed products for feed and food industries. *In* Alfalfa science and technology. Agronomy 15: 660–676. Am. Soc. of Agron., Madison, WI.

Langston, C. W., H. Irvin, and C. H. Gordon, C. Bouma, H. G. Wiseman, C. G. Melin, L. A. Moore, and J. R. McCalmont. 1958. Microbiology and chemistry of grass silage. USDA Tech. Bull. 1187.

Lechtenberg, V. L., K. S. Hendrix, D. C. Petritz, and S. D. Parsons. 1979. Compositional changes and losses in large hay bales during outside storage. p. 11–14. *In* 1979 Purdue Cow/Calf Research Day, 5 April 1979. Purdue Univ., W. Lafayette, IN.

Longhouse, A. D. 1960. Hay conditioners in the Northeast United States. West Virginia Agric. Exp. Stn. Bull. 449.

Minson, D. J., and R. J. Lancaster. 1965. The efficiency of six methods of covering silage. N.Z. J. Agric. Res. 8:542–554.

Moore, L. A. 1951. New values in good hay and silage for dairy cows. Bur. Dairy Ind. Inf. Mimeo 117.

Owen, F. G. 1971. Silage additives and their influence on silage fermentation. p. 79–112. *In* Technological papers presented at the International Silage and Research Conference. 6–8 Dec. 1971.

Parsons, S. D., V. L. Lechtenberg, D. C. Petritz, and W. H. Smith. 1977. Storage and feeding of big package hay. p. 290–292. *In* Int. Grain and Forage Harvesting Conference. Ames, Iowa, September, 1977. Am. Soc. Agric. Engineers Publication 1-78. St. Joseph, MI.

Renoll, E. S., W. B. Anthony, L. A. Smith, and J. L. Stallings. 1974. Stack and bale systems for h ay handling and feeding. Auburn Univ., Alabama Agric. Exp. Stn. Bull. 455.

Rounds, W., J. Waller, and T. J. Klopfenstein. 1975. Chemically treated crop residues. p. 9–10. *In* 1975 Nebraska Agric. Exp. Stn. Beef Cattle Report. EC 75-218.

Shepherd, J. B., C. H. Gordon, and L. E. Campbell. 1953. Developments and problems in making grass silage. USDA Bur. Dairy Ind. Inf. Mimeo 149.

Shrock, M. D., and G. E. Fairbanks. 1975. A comparative study of harvesting efficiency for three types of packaging machinery on alfalfa. Paper MC-75-902. Mid-Central Region Meeting, 22 March 1975. Am. Soc. Agric. Engineers, St. Joseph, MI.

Smith, D. 1969. Removing and analyzing total non-structural carbohydrates from plant tissue. Wisconsin Agric. Exp. Stn. Res. Rep. 41.

Smith, K. D. 1977. What is coming in alfalfa dehydration. *In* Alfalfa: green gold in the Great Plains. p. 17–18. Proc. Seventh Annu. Alfalfa Symposium. 22–23 March 1977. Sioux Falls, SD.

Smith, W. H., V. L. Lechtenberg, S. D. Parsons, D. C. Petritz. 1974. Suggestions for the storage and feeding of big package hay. ID-97. Purdue Univ. Cooperative Ext. Serv., W. Lafayette, IN.

Van Soest, P. J. 1965. Use of detergents in analysis of fibrous feeds. III. Study of affects of heating and drying on yield of fiber and lignin in forages. J. Assoc. Off. Agric. Chem. 48:785–790.

Waiss, A. C., J. Guggolz, G. O. Kohler, H. G. Walker, and W. N. Garrett. 1972. Improving digestibility of straws for ruminant feeds by aqueous ammonia. J. Anim. Sci. 35:109–112.

Whittenbury, R., P. McDonald, and D. G. Bryan-Jones. 1967. A short review of some biochemical and microbiological aspects of ensilage. J. Sci. Food Agric. 18:441–444.

SUGGESTED READING

Dale, J. G., D. A. Holt, R. M. Peart. 1978. A model of alfalfa harvest and loss. Paper No. 78-5030. Summer Meeting Am. Soc. Agric. Engineers, 27–30 June 1978. Logan, UT. Am.Soc. Agric. Engineers, St. Joseph, MI.

Hanson, C. H. (ed.). 1972. Alfalfa science and technology. Agronomy 15: Chapters 12, 26, 27, and 30. Am. Soc. Agron., Madison, WI.

Heath, M. E., D. S. Metcalfe, and R. F. Barnes. 1973. Forages. Chapters 48 to 52. Iowa State Univ. Press, Ames.

Hundtoft, E. B. 1965a. Handling hay crops—capacity, quality, losses, power, cost. Cornell Univ. Agric. Eng. Ext. Bull. 363.

————. 1965b. Handling hay crops—from standing crop to windrow. Cornell Univ. Agric. Eng. Ext. Bull. 364. Ithaca, NY.

Langston, C. W., H. Irvin, C. H. Gordon, C. Bouma, H. G. Wiseman, C. G. Melin, L. A. Moore, and J. R. McCalmont. 1958. Microbiology and chemistry of grass silage. USDA Tech. Bull. 1187.

Musgrave, R. B., and W. K. Kennedy. 1950. Preservation and storage of forage crops. Adv. Agron. 2:273–315.

Owen, F. G. 1971. Silage additives and their influence on silage fermentation. p. 79–112. *In* Technological papers presented at the International Silage and Research Conference, 6–8 Dec. 1971.

Sullivan, J. T. 1969. Chemical composition of forages with reference to the needs of the grazing animal. USDA Rep. ARS 34-107.

Watson, S. J., and M. J. Nash. 1960. The conservation of grass and forage crops. Oliver and Boyd, Edinburgh and London.

Whittenbury, R., P. McDonald, and D. G. Bryan-Jones. 1967. A short review of some biochemical and microbiological aspects of ensilage. J. Sci. Food Agric. 18:441–444.

INDEX

Q

R